Discovering Chemistry
Semester 1

Student Manual

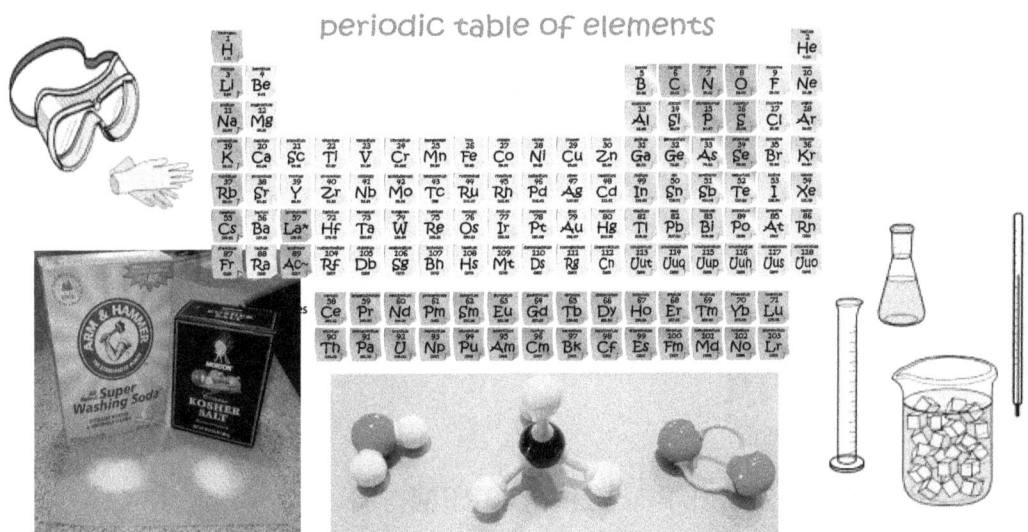

periodic table of elements

By

Frankie Wood-Black, Ph.D., REM, MBA

Discovering Chemistry - Semester 1
Student Manual

Written By: Frankie Wood-Black, Ph.D., REM, MBA
Copyright 2015 Sophic Pursuits, Inc.

Publisher's Cataloging-in-Publication data

Wood-Black, Frankie, 1963-
         Discovering Chemistry - Semester 1 - Student Manual

ISBN 978-1-940843-03-2 (Printed Version)
1. Science - Chemistry - General 2. Science - Experiments & Projects -
Chemistry

Dedication

Thank you to my husband and children who were the first testers of these experiements. Thank you to the homeschool families that piloted the material. Thank you to all my general chemistry students that have helped refine the information.

With out your help and suggestions, this course would never have been produced.

Disclaimers

While the author has used every effort to provide materials that are safe for students and their instructors, the author cannot anticipate exactly how the material will be used. The author has utilized accepted references, consulted with individuals with experience in health and safety related matters, and has conducted the individual activities in both a laboratory and a home setting. The materials chosen for this course are readily available and should be familiar to most individuals using the course materials. Additionally, the author has attempted to include good laboratory practice techniques and safety information for all of the laboratories to be performed.

Each individual learning situation will be different and the potential hazards associated with specific laboratories presented may or may not have been addressed by the author as the specific setting and conditions under which the laboratories will be performed are not under the control of the author. Thus, the author has made recommendations throughout the text and anticipates that the laboratories will be conducted under proper supervision by a parent or instructor.

The author is not making any claims, warranties, or guarantees related to the sufficiency of the information contained in this document. As the specific situation and/or environment cannot be anticipated by the author, this document in no way outlines all the necessary warnings and precautionary measures which may need to be taken prior to conducting a hands-on activity. Nor, does this document meet or comply with the requirements of any safety code or regulation.

Users of this document should consult specific safety information provided with the materials, instructions, and/or other reference materials to determine hazards associated with each individual laboratory as recommended in the text. Warnings and precautions should be reviewed and evaluated prior to conducting any laboratory.

Contents

Notes to the Student

Using this Course

This course is a stand-alone laboratory course designed to be used by students that do not have access to a laboratory. It is designed to be used in conjunction with a traditional general chemistry course. Thus, this course can be used for home school or online course applications.

The materials for this course were chosen based on their minimal risk potential and availability. The experiments utilize typical household materials and products that can be easily found at a local grocery or hardware store. Some items may be obtained using online sources.

Each laboratory as presented, provides an introduction, safety considerations, procedures, and discussion. Prior to beginning each laboratory, you should read the laboratory completely to understand the basic material being explored, the safety considerations, the equipment needed and how the experiment will proceed.

The laboratories were chosen to help develop critical laboratory skills. As such the difficulty and expectations of the laboratories increase as you move through the course. As with any typical science class, the material builds upon concepts learned earlier, so it is important that you do the laboratories in order.

Laboratory # 1
Exploration of Chemicals and Chemical Information
Chemical Information Scavenger Hunt
Focus on Safety Information

<u>Introduction</u>

Chemistry is considered the "central science." This is because chemistry connects all of the other sciences. For example, think of our universe as the sum of various physical laws such as – energy is neither created nor destroyed, an object in motion remains in motion, etc. Chemistry builds upon these laws. Atoms and molecules interact based upon these physical concepts. Ultimately, we begin to look at how atoms and molecules interact in cells or how animals convert sugars into energy. Thus, chemistry is an essential building block and the science around which we can explain a number of things that are in our world.

Everything is made up of chemicals. The air we breathe is a mixture of small **molecules** like carbon dioxide, oxygen, nitrogen, and a number of other things. We can't see the individual molecules but we know they are there. Water is a chemical. It is made up of atoms of hydrogen and oxygen. Your soda pop is a mixture of chemicals – sugars, colorings, carbon dioxide, and flavorings. Chemicals are all around us.

Yet, when we hear the word chemical, we tend to think of something negative or bad. Why? Because there are some chemicals that are toxic to people, animals, plants, etc. Think about a nice lawn. In that nice lawn there are grasses that we want to have, but every now and then a weed (i.e. a plant we don't want in the grass) creeps in and spoils our lovely uniform lawn. **Pollution** is much like this. Air pollution is just chemicals that we don't want, like carbon monoxide. Similarly, water pollution can be described in much the same way.

We know there are materials that are considered **poisons**, i.e. are harmful to human health if consumed or absorbed in the wrong amounts. Arsenic and cyanide are common ones. But, did you know that many vitamins and minerals can also become poisons if we **ingest** them in the wrong amounts? It is very important that we understand the chemicals that are in our surroundings. It is important to know how to properly handle, store and use them.

In order to safely handle these materials, a person must be able to identify the **hazards** associated with them and be able to reduce those hazards. For example, if the material that we are using is vinegar (acetic acid) we know that this material can react with certain other chemicals (baking soda - sodium bicarbonate). We know that the material can give off vapors which can irritate the eyes and nose. Vinegar, which is in our kitchen, is a dilute form of acetic acide and the hazards while irritating are not likely to be very dangerous. Let's look at a different common material, gasoline. Gasoline is flammable and can be easily ignited. It is harmful if swallowed and has the potential to have long-term health implications if a person is over-exposed to it on a daily basis. It has the potential to be very dangerous.

This laboratory is more of an activity. Laboratories can either be **experiments** or **activities**. For the purpose of this course an experiment is defined as:

> *A procedure carried out with the expressed goal of verifying, refuting or establishing the validity of a hypothesis.*

As we are not testing a specific scientific principle or concept here, we are not conducting an experiment. This is an activity designed to help you "discover" the various types of chemicals that can be found in your own home and how to find information about these materials. The purpose of the activity is to familiarize yourself with safety information about the materials and how you can use this information to **mitigate** potential hazards.

During the course of the activity, you will be identifying safety equipment and the concept of a hazard analysis will be introduced.

General Safety Considerations for this Activity

During this activity you will be handling a number of chemicals. Although, these chemicals may be routinely found in the home or garage, you will need to use caution when handling materials. For the purpose of this activity, no chemicals or materials should be removed from the packaging.

Some cautions –

- Do not remove materials from the packaging.

- If residues are on the outside of the packaging – do not handle without adult supervision.

- Wash hands at the completion of this exercise.

Objectives

- Identify common materials
- Understand chemical labels
- Locate material safety information
- Identify safety equipment
- Define a hazard analysis

Procedure

This is a material and information scavenger hunt. For this activity, you will need the following items:

- Activity Worksheet
- Internet Access

Step 1 – Finding the focus materials

Find an example of each of the items listed below. Items need to be in their original packaging, i.e. the packaging from the manufacturer, and should be legible.
For each item you will need to be able to identify:

- The manufacturer
- The suggested use(s)
- Any warnings or cautions
- A contact phone number

Items to be found:

- A detergent for cleaning dishes
- A toilet cleaner
- A fabric softener
- A fertilizer

Complete Step 1 of the data collection sheet.

Discussion of Step 1

Per governmental regulations in the United States (note there are similar regulations in Europe as well as other countries), consumer products are required to be labeled with certain information. Each label must meet the specific guidelines as outlined in the regulations. Under these regulations, a consumer product is defined to be hazardous if:

The material is toxic, corrosive, an irritant, a strong sensitizer, flammable or combustible, or generates pressure through decomposition, heat or other means, and may cause personal injury or substantial illness during or proximate result of any customary or reasonably foreseeable handling or use or ingestion by children.

For these hazardous materials, the label must include:

- Signal words like WARNING, CAUTION, DANGER, and POISON
- Affirmative statements such as Harmful if Swallowed or May Cause Irritation
- Name and place of business of the manufacturer, packer, distributor or seller
- Common or usual name or chemical name
- Precautionary measures to follow
- Instructions for first aid treatment when appropriate
- Instructions for handling and storage
- "Keep out of the reach of children" or its practical equivalent

Additional information may be required due to the specific product or use. These requirements are included and referenced in the governing regulations.

The idea behind such labeling laws is to provide the user of the material information about the potential hazards and how to use the product safely. It is important to read the label prior to using the product even if you are familiar with it and its use as the manufacturers are constantly improving and updating information.

During Step 1, you captured information about four products. Answer the following questions regarding your specific products:

1. What signal words were present?

2. Were there any affirmative statements like Harmful if Swallowed? If so, what were they?

3. Were there specific handling requirements identified? For example: Use in ventilated area, or wear gloves when handling. If so, what were they?

4. Was a recommended first aid treatment listed? If so, what treatment(s) was(were) listed?

5. What precautionary measures were listed?

Step 2 – Finding additional information for the focus materials

For this part of the activity you will need to have Internet access.

In addition to the product label, consumer products are required to have additional safety and/or chemical information. The most common piece of information is the Safety Data Sheet. For most consumer products, the Safety Data Sheet can be found on the Internet or can be obtained by calling the manufacturer.

The Safety Data Sheet contains more detailed information about the product. It contains additional physical and chemical properties. This is helpful information that can't be included on the small space that is the label or other packaging.

Using the Internet, see if you can find a Safety Data Sheet for all of the materials you found in Step 1. For the rest of this step and Step 3, pick one Safety Data Sheet for further analysis. Complete Step 2 on the Data Collection Sheet.

There has been a recent change in the layout of the Safety Data Sheet to allow it to meet the requirements of the Global Harmonization Standard (GHS). Prior to the implementation of the GHS regulations beginning in 2014, the Safety Data Sheet (SDS) was called the Material Safety Data Sheet (MSDS). During this activity you may find both the MSDS and/or the SDS.

Discussion of Step 2

The SDS is intended to provide the user of the product with specific information regarding the potential hazards, ingredients, regulatory data, physical properties, and reactivity. These SDSs came into existence because individuals have a "right-to-know" what potential hazards they may be exposed to by using the product and how to use it safely. The label on the bottle or box can only provide a small amount of information due to the limited space. The SDS can provide more information because it is not limited to size of the container.

Working through the activity on the Data Collection Sheet, the specific ingredients with the chemical name, hazards, health, reactivity and handling information was recorded. By reading the SDS, anyone who uses the product should be better informed about its potential dangers and how to safely handle and dispose of any unused product.

Let's look at the information a bit more closely:

In the first part of Step 2, the product is identified both by the common product name and a listing of the hazardous ingredients. The hazardous ingredients are listed by name and by CAS number. The CAS, Chemical Abstract Service, number is a registry number assigned by the American Chemical Society Chemical Abstract Service. This number is intended to uniquely identify the chemical compound for ease of communication between scientists, regulators, and chemical users. There are other product registries and the SDS may list them in other sections.

In the identification of the hazardous ingredients, the concentrations and worker exposure limits are also provided. (If the worker exposure limits are not provided next to the ingredients, they are generally included in another section of the SDS.) These limits when fully understood inform workers about the safe limits of using that specific chemical without any adverse health effects.

After identifying the hazardous materials, information about who to contact in case of emergency and how to handle the material is gathered. In these sections, information about protective clothing will be presented. Information about potentially harmful reactions and health effects are also listed or discussed.

In the SDS, recommended first aid treatments are provided to aid individuals adversely impacted by the material and can be used by emergency workers responding to an accident involving the product. Should an individual become injured when using the product, this information is essential so that any treatment provided does not continue or further harm the exposed individual.

Finally, how to properly dispose of the product is recorded. This information is provided so that the product does not harm the environment or other workers. Some materials are not allowed in municipal trash and must be disposed of on special pick-up days or at a specific waste center.

Step 3 – Completing the hazard analysis

Given the information obtained in Step 2, a hazard analysis can be conducted for using the selected product. The idea behind the hazard analysis is to help a person stay safe when conducting an activity. The purpose of the hazard analysis is to think about potentially bad situations that might occur during the normal use of the product. Having identified those potentially bad situations, you can figure out ways to prevent them from happening.

It is also important to talk about another safety concept – risk. Risk is defined as the possibility that a negative situation may occur. For example, there is a risk that you might fall when you go up or down stairs. Risk is the likelihood that this may occur. Your risk of falling is reduced if you use the hand rails. The same goes for the use of materials and/or products.

From the SDS, hazards were identified. Look at the stair example – a hazard might be a loose rug or a toy on the stairs. This hazard may increase the risk of falling. Hazards and risks are not the same, but hazards can increase the risk. For our stair example, the hazard could be fixed or removed. Therefore, the risk will be dimiished or eliminated.

So, let's look at one form of a hazard analysis (the last page of the worksheet). For our hazard analysis, we are going to look at using the product that you used to complete Step 2. For our example, we are going to use a window cleaner and the activity of cleaning a window.

Look at the first question: "Are there any hazards associated with the materials being used?" For our task, cleaning a window, we are going to assume that the materials being used are the window cleaner and a paper towel. So, in that first section we would list the window cleaner. (We are going to assume that there is no hazard associated with the paper towel.) Then, we ask the question are there any hazards or warnings associated with the window cleaner. Our window cleaner has ammonia in it and there is a warning on the label. So, we would list ammonia as a hazard – and put the warnings listed on the label into our table.

From the SDS, there may or may not be any additional hazards. But, for our window cleaner there may be a warning either on the label or the SDS that says DO NOT MIX WITH BLEACH. Ammonia and bleach will reach to form a strong base (Sodium Hydroxide, NaOH), dichloramine ($NHCl_2$), and nitrogen trichloride (NCl_3), and a toxic gas. The SDS will definitely provide more information about the reaction with bleach. So, you now can complete the first table.

Once, the initial table is complete. Begin to answer the questions. "What adverse reaction and/or event might occur when using this material?" "What are the potential bad outcomes?" "What safety equipment should be used?"

For our window washing activity (we are doing this instead of an experiment), think about the potential things that can go wrong. We could –

- Spill the material
- Mistakenly mix it with bleach
- Get it on our hands

So, we can complete the answer to that first question and list the bad outcomes.

Now, think about what you can do to prevent these things from happening and what safety equipment you might use. Looking at our example:

•Spill the material – To prevent, make sure the cap is on the bottle.

• Mistakenly mix it with bleach – To prevent, keep the window cleaner away from bleach and do not use when bleach has been used.

• Get it on our hands – To prevent, wear gloves.

You can now check the boxes in the last section.

Now, you are ready to complete this for your selected product.

Discussion of Step 3

The purpose of the Hazard Analysis is to help you to prepare for the potential bad outcomes. This is much like a fire drill. You want to be ready in case there is a fire, first by preventing the fire and second by knowing what to do in the event of a fire. The Hazard Analysis gets you to think about what you are doing before you do it and helps you to prepare just in case the spill or accident happens.

This form of the Hazard Analysis has been prepared to help you throughout the rest of this course. You will be asked to complete the Hazard Analysis as part of each experiment. (There will be some activities where this will not be necessary and these will be noted as we go through the course.) This will get you into the habit of thinking about what you are about to do before you do it, thereby making the process safer and accidents less likely to happen.

A couple of other things about safety. Have you ever had that little voice in your head say something that sounded like:

"If I pull that plate out from the bottom, the other plates might fall?"

"I really should get a step ladder, and not climb on the counter to get that?"

"I shouldn't leave that out on the counter, because…"

All of these are little warnings that something bad may happen. The basic idea behind safety is that if you have these types of thoughts, listen to them. They are warnings. Yes, you may get away with it this time without having something bad happening but eventually something bad will.

Safety is also a team sport. You because of your experience or perspective may not necessarily see all of the potential hazards associated with the experiment or activity. Having someone around to discuss or be a second set of eyes is always a good idea.

*** *Keep a copy of all of the SDSs you use. You will use some of the information in later activities/experiments.* ***

Chemical Information Scavenger Hunt

Data Collection Sheet

Step 1

Find the following items:
- A detergent for cleaning dishes
- A toilet cleaner
- A fabric softener
- A fertilizer (or a garage product)

Item 1 - Detergent	Item 2 - Toilet Cleaner
Name: _____	Name: _____
Manufacturer: _____	Manufacturer: _____
Suggested Uses from Label:	Suggested Uses from Label:
_____	_____
Warnings and/or Cautions	Warnings and/or Cautions
_____	_____
_____	_____
_____	_____
Contact Information (Phone No. and/or Website)	Contact Information (Phone No. and/or Website)
_____	_____

11

Item 3 - Fabric Softener

Name: _____

Manufacturer: _____

Suggested Uses from Label:

_____

Warnings and/or Cautions

_____

_____

_____

Contact Information (Phone No. and/or Website)

_____

Item 4 - Fertilizer (or Garage Product)

Name: _____

Manufacturer: _____

Suggested Uses from Label:

_____

Warnings and/or Cautions

_____

_____

_____

Contact Information (Phone No. and/or Website)

_____

Step 2

Pick one of the Items from Step 1 - Conduct an Internet search to find the Safety Data Sheet (SDS) for that material. Complete the following table:

Name of the Material: _____

Internet Location of the SDS: _____

List the Hazardous Ingredients with CAS#:

_____

_____

Emergency Contact Information:

_____

Special Protection and Precautions (Recommended Handling Practices)

_____

_____

_____

_____

Is the material reactive? i.e. Are there any dangers associated with mixing the product with other materials or heating or cooling the product?

_____ Yes   _____ No

If yes - what materials should not be mixed with the product or what conditions should be avoided:

_____

_____

_____

_____

What health hazards are identified?

_____

_____

_____

_____

If the material is not used completely how should the unused product be disposed?

_____

_____

_____

How many sections where included in the SDS and list:

Number of Sections: _____

_____

_____

_____

_____

_____

_____

_____

_____

_____

_____

_____

Hazard Analysis

Are there any hazards associated with the materials being used?

	Material to be Used	Warnings/Hazards Known	Warnings/Hazards Known from Label	Warnings/Hazards Known from Safety Data Sheet or other Source
1.				
2.				
3.				
4.				
5.				

What adverse reaction and/or event might occur when using this material?

___ Spills   ___ Fire   ___ Explosion   ___ Heat

___ Cold   ___ Solids   ___ Projectiles   ___ Loud Noise

___ Other: _____

___ Other: _____

What are the potential bad outcomes from the experiment if something goes wrong?

_____

_____

_____

What safety equipment should be used based upon the hazards of the materials and the anticipated behavior and/or outcome of the activity?

___ Splash Goggles   ___ Safety Glasses   ___ Gloves – Type: _____

___ Science Smock   ___ Fire Extinguisher   ___ Barricade/Perimeter – Type: _____

___ Spill Clean-up   ___ First Aid Kit   ___ Other: _____

___ Other: _____

Laboratory #2
Exploration of Chemicals and Chemical Information
Chemical Scavenger Hunt
Finding Chemicals Around You

Introduction

Chemicals are all around us, and we are all a product of chemistry. In fact, the American Chemical Society has a sticker set and game that is used for National Chemistry week where the challenge is to find something that does not contain chemicals. (This is a trick question, of course, in that everything is made up of atoms which is the basic unit of a chemical element.) When you understand this, you begin to realize all of the misinformation that is out there in the world of advertising. How can they really say that this water has no chemicals or that all of the chemicals have been removed from the water? Water is a chemical compound made from hydrogen and oxygen. It has the chemical formula H_2O, two atoms of hydrogen for every one atom of oxygen.

Everything we see, feel, taste and smell is made from chemicals. Some of the chemical formulas are simple like water and table salt (the formula for table salt is NaCl, sodium chloride). Others may be very complex in terms of the number of atoms and/or elements they contain. Examples of complex molecules are sugars, starches, proteins, and DNA (deoxyribonucleic acid). Moving up from single molecules to mixtures or products, the complexity of the interactions increases. Think of a bottle of shampoo. Shampoos contain water, detergents, and perfumes. Just how many different chemicals and the types of interactions may be present?

In our first activity, the chemical safety scavenger hunt, we looked at common household products and learned where to find important information about how to use the product and keep us safe. We looked at the labels, and the safety data sheet (SDS). And, hopefully, you found that there was chemical information: the chemical formula and its CAS number. (The CAS is the chemical abstract number that has been assigned by the American Chemical Society to help uniquely identify the chemical substance.) You may have also found a molecular structure.

In this activity, your mission is to identify many chemicals that are commonly used around your house. People are performing chemistry at home, every day. Think about the many activities that require chemistry? Washing your hands – you use a detergent to remove the greases and oils. Have you ever used a carpet stain lifter?

Or, how about removing rust? How do you get your gravy to thicken? All of these are chemical reactions. But, how do you know what chemicals are being used?

There is one more thing about chemicals – most of the common chemicals go by their common names. Table salt, for example, is really sodium chloride, NaCl. Finding the true chemical name and molecular (chemical) formula may require going to more than one source. The ingredients with the chemical name will generally be listed in the ingredients section of the SDS. Once you have the name, you can do a search for the chemical formula using either the name or the CAS#. The CAS# is the Chemical Abstract Registry Number used by the American Chemical Society to uniquely identify the chemical.

So, now it is time to go on a scavenger hunt and find out just "what is that stuff" on a molecular level!

> **The molecular or chemical formula is symbolic representation of the chemical using the symbols for the atoms us contained in the molecule. The formula includes the number of each atom used in the molecule. Recall the example of water, H_2O. It contains 2 atoms of hydrogen and 1 atom of oxygen in each water molecule.**

General Safety Considerations for this Activity

As was mentioned in the first activity, we will be doing hazard analyses for all of our experiments. This is an activity – there is a difference between an experiment and an activity. In an experiment, you are testing a hypothesis and the outcomes are not necessarily certain. In an activity, you are primarily performing a task. Many tasks will require you to conduct a hazard analysis as well; however, for this activity, you are primarily looking at labels, and obtaining information from the Internet or other sources. This activity does not require a hazard analysis as discussed as the physical risks associated with doing some Internet research does not typically involve activities which have significant "bad outcomes", or could cause you harm.

Yet, if you think about it – this activity does pose some risk. For example: if you store your spices in a cabinet that is difficult for you to reach from the ground, getting that container, may pose a hazard (you could fall). Some the materials that you may be looking at in the garage or under the sink may be hazardous, antifreeze and cleaning products contain materials that require special precautions; as you learned in the last activity. So, you still need to use caution.

Just because a chemical is common or can be readily purchased, does not mean that there are no hazards associated with them. You still need to pay attention to warning labels. You need to take care when getting things from their storage areas. You need to be aware of your surroundings. So, keep the following "rules" in mind when conducting this activity:

- Make sure someone else is aware of what you are doing

- Follow any household or classroom guidelines associated with using the Internet and/or other reference materials

- Be aware of your surroundings, look for hazards: tripping, heights, other containers, etc.

- Read the labels, be aware of any cautions or warning statements

For this activity, please do not open the packages or handle any of the material directly. Use your familiarity with the product or item to answer the questions.

Equipment required for the Activity

- Access to reference materials: Internet or other materials
- Notebook

Objectives

- Identify common household items by a variety of names: common, chemical, and molecular (chemical) formula – example: Bleach (common name), Clorox™ (trade name), sodium hypochlorite (chemical name), and NaClO (chemical formula)

- Establish a laboratory notebook and document the laboratory activities.

- Recognize chemicals around you

- Compare and contrast the different chemicals

- Identify compounds (molecules)

- Identify pure compounds versus mixtures

- Find the molecular formula for common materials

Activity/Procedure

This activity is a good introduction to taking notes. As mentioned in the student introduction, during this course you are going to be developing laboratory skills. One of those skills is the recording of information about your activities and findings. In this activity, you will be collecting a variety of information about household products and be using a variety of sources. This information will be collected and documented in your laboratory notebook.

Note taking is a skill used by all scientists. The notes taken by the scientist cover a variety of topics, everything from the references used, general information, experimental conditions, observations, ideas, hypotheses, experimental results, initial conclusions, etc. Scientists use this information to develop new experiments and new hypotheses. The laboratory notebook becomes an essential tool and record for discoveries. They are used to document findings, establish records of invention, and provide the detail of the experimental conditions and tests.

The challenge, of course, is coming up with a systematic method of note taking and documentation in the notebook so that this information can be used not only by you, but by someone else. Being able to reproduce your results is critical to the scientific method and process. The only way for you to learn what works for you is through practice and repetition. Throughout this course you are going to get multiple opportunities to work on your note taking skills. While activity sheets and data collection sheets may be provided as part of the materials in each laboatory; as you move through the course you will rely more on your notes and notebook. This activity is your first opportunity to get started on this task.

<u>First, what is the activity?</u>

During this activity, you are going to obtain chemical information for the following items:

- Baking Soda
- Washing Soda
- Ammonia
- Antifreeze
- Fertilizer
- Vanilla
- Vinegar

For each of the items, you will need to obtain:

- The common name
- The trade name or product specific name
- Chemical name
- Chemical formula
- What the material is used for or common use
- Warnings/Cautions associated with the material
- Any special handling instructions
- Emergency contact information
- Physical state of the material (solid, liquid, gas)
- Description of the product (physical observation)
- Any other information that you think might be relevant.

Setting up your laboratory notebook

Now that you know what you are going to be doing and the information that you are going to be collecting you can begin to set up your notebook. Remember that your notebook is going to be documentation for this laboratory course, so you will need to set it up that way. The first page should be the title page and should look something like:

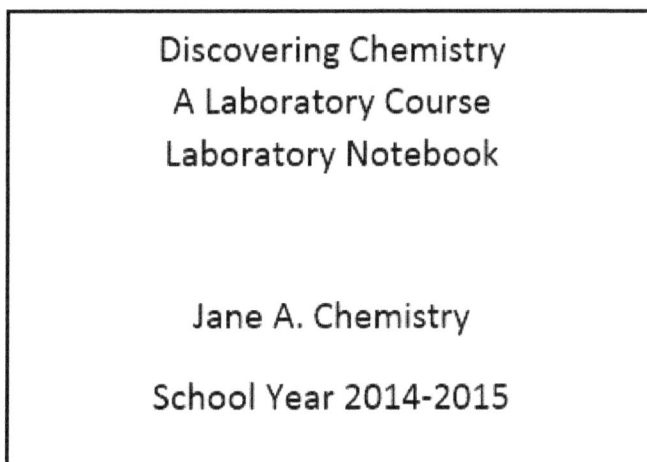

```
Discovering Chemistry
A Laboratory Course
Laboratory Notebook


Jane A. Chemistry

School Year 2014-2015
```

Leave the next two pages blank. These will become your table of contents. (Note: it is a good idea to number your pages, front and back. (This allows someone to know that you haven't ripped or torn any out.)) So, starting at about page 5; begin to lay out your activity. Using bleach as an example, your page may look something like this:

Activity – Chemical Information Scavenger Hunt

Objective: Find information about 7 common household chemicals

Information to be obtained includes:

Common name, chemical name, chemical formula, safety information, physical information, etc.

Chemical 1:

Common Name: Bleach

Trade Name: Clorox Regular Bleach

Chemical Name: sodium hypochlorite

Chemical Formula: NaClO

CAS#:   7681-52-9

Physical description: Liquid – light yellow with a characteristic chlorine odor

Uses: Household cleaning and bleaching of clothes in laundry

Cautions/Warnings: Danger, Corrosive – may cause irritation or damage to eyes and skin. Vapor or mist may irritate. Harmful is swallowed. Keep out of reach of children.

Reactivity: Stable under normal use and storage. REACTS with other household chemicals such as toilet cleaners, rust removers, vinegar, acids or ammonia to produce a hazardous gas.

Special Precautions: Avoid contact with eyes, skin and clothing.

Emergency Contact Information: Medical – 1-800-446-1014

Other Information: Generally found in concentrations of 5 to 10% in household product.

Household product may also contain sodium hydroxide (NaOH) at <1%

Source of Information: SDS from Clorox Company on the internet, copy of SDS is on computer hard drive.

Initials: *JAC*

*** *Keep a copy of all of the SDSs you use. You will use some of the information in later activities/experiments.* ***

Safety aside – one of the comments about bleach is its characteristic odor. When working with hazardous chemicals, a person avoids contact with the chemical, i.e. you do not want to have the material come into contact with the skin or ingested. Using your sense of smell becomes a bit tricky, as many chemicals are dangerous when inhaled. Yet, smell is also used for identification of certain chemicals. Thus, there is a laboratory technique called wafting.

Wafting is done by moving a vial of the material gently back and forth so that the vapor or smell "floats" in the air and you can smell it without being overcome. As with any laboratory course, you should not directly smell the chemicals unless instructed to do so and then only by wafting.

For highly odorous or where concentrations of vapor may cause a hazard, chemists use other protective measures to ensure that exposure to the vapors is eliminated or significantly reduced. Safety equipment such as laboratory hoods, and respirators are used in these cases. This course uses materials that are commonly found, thus these extra safety precautions will not be required.

You can see from this laboratory notebook page, some information was added above what was requested in the list of essential information. This was included because JAC thought that they might need that information down the road. Also, the source of the information was documented. Even though that was not requested in the list of information; this item should be recorded for all reference material. Where did I get this piece of information? This is important as it documents and gives credit to the appropriate source.

JAC could have taped or stapled the SDS into the notebook as well. But, since this was a reference from the Internet, recording the website, or location of the SDS is just as acceptable. As you move through this course, you will want to tape graphs or other information into your notebook. Anything that you add to your notebook should include your initials and each page should be initialed once it is complete.

Continue collecting information on the other six common items. Some things to think about while you are collecting the information. Are the chemicals simple or complex? For bleach, the active ingredient is the sodium hypochlorite. This is a fairly simple compound made up of three different elements. As noted the product is a mixture containing 5 to 10% sodium hypochlorite with the remainder being water. The mixture is a homogeneous one, as the mixture is uniform throughout. (Another type of mixture is the heterogeneous mixture. A heterogeneous mixture is not uniform throughout. In the heterogeneous mixtures one sample of the mixture may have a different concentration than another sample. Concrete, soils, and vegetable soup are all examples of heterogeneous mixtures.) But, think about some of the other products, are the chemical formulas going to be as simple? Or, more complex?

Our consumer products are typically mixtures. What is the carrier? For example, flavorings may be a mixture of the volatile oil (like peppermint oil) and alcohol. Even solids can be mixtures. Look at ground pepper. You can have multiple types of pepper corns ground together. A gravy mix has corn starch, flour, and dehydrated beef broth. The flour can be considered the carrier for the gravy mix.

Is there more than one active ingredient? Is there an inert (meaning that the ingredient is there for color or packaging and does not participate in the product) ingredient? For example: many powdered products contain ingredients that help to prevent clumping allowing for easy pouring.

Summing It Up:

While this activity was designed for you to get familiar with note taking and finding information on the various chemical compounds that are present in the household, there are some other things that you may have noted.

- As you went through the activity, were you able to predict whether or not the chemicals were complex or simple?

- Did you happen to compare the same common product from one or more manufacturers? Clorox is not the only manufacturer of bleach. Were the ingredients the same from one manufacture to another? Did they have the same types of information?

- Was the molecular (chemical) formula hard or easy to find? Did you have to go to more than one source?

- Were you able to distinguish between a homogeneous and heterogeneous mixture? Did you have an example of a heterogeneous mixture in one of your items?

- As you went through the activity, did you tend to gather more information about your item?

Let's return to the initial objectives of the activity:

- Identify common household items by a variety of names: common, chemical, and molecular (chemical) formula – example: Bleach (common name), Clorox™ (trade name), sodium hypochlorite (chemical name), and NaClO (chemical formula)

- Establish a laboratory notebook and document the laboratory activities.

- Recognize chemicals around you

- Compare and contrast the different chemicals

- Identify compounds (molecules)

- Identify pure compounds versus mixtures

- Find the molecular formula for common materials

Did the activity meet all of these objectives?

Take some time to summarize your thoughts about this activity in your laboratory notebook. This can be done in a discussion or observation section in your notebook.

Laboratory #3
Understanding Measurements
Accuracy and Precision

Introduction

In all of the sciences, there are some key experimental concepts that must be understood in order to properly evaluate experiments. These three concepts are related to the quality of the experimental data being collected. Are the values that you have collected correct?

Randomness or chaos is a part of the world around us. Think about rolling a pair of dice. The values that may be obtained range from two (a one on each die) to twelve (a six on each die). But, unless a die is weighted or modified – the chance that a one, two, three, four, five or six will appear is left to chance. This means there is an equal likelihood that each of the numbers may appear on top. If the die is weighted or modified, this means that you have artificially changed the die so that a specific number is more likely than any other number – the die is said to be biased. Bias means that there is some specific reason that one value over another is likely to occur. We see bias in a number of ways – the weighted die is one example, but improperly calibrated instruments such as a bathroom scale or a thermometer could also result in bias.

The idea that randomness occurs in nature allows us to determine the error – the difference between the real or true value and what we have determined experimentally. If there is no bias, it is likely that we are going to measure the value higher or lower equally over the course of many trials. For example, let's say that the true value of a weight is 100.0 grams. We measure it ten times on a scale that allows us to read the weight to 0.0 grams. The ten readings might look like this:

100.2 grams	100.3 grams
99.9 grams	100.0 grams
100.0 grams	99.8 grams
100.1 grams	99.7 grams
99.9 grams	100.1 grams

In these ten readings we have two that are the exact weight and an equal amount that area above and below the actual weight. Of course, this was a specific set of data and it is likely that your own data may vary a bit; due to the natural randomness of nature.

If we average the values listed we see that the value obtained is 100.0 grams – which is exactly what the true value was intended to be. Thus, this value has essentially no error and is deemed to be accurate. Accuracy is the how close the measured value is to the accepted value.

We may also look at the data and determine whether or not our 10 measurements were precise. Precision measures how close the individual measurements are to each other. In the case of our ten measurements, the least precise measures are the values of 100.3 and 99.7 as they are further away from the true value.

You can have measurements that are precise, i.e. close together, but not accurate. And, you can have measurements that are accurate but not precise. We can demonstrate these scenarios by looking at the measurements as if they were darts on a dart board and the center of the dart board is the true value. (See Figure) In Figure 1, you can see that a measurement can be accurate and precise, and combinations of the others.

When a value is precise and deviates from the true value – you can examine the type of bias that may be present.

During this activity, you are going to explore how you take the measurement might impact the accuracy and/or the precision. During this experiment, you are going to determine the weight of a penny and compare it to the United States Mint's specifications. You are going to conduct the experiment two different ways and compare the methods as well as your results. Additionally, you are going to look for sources of error.

General Safety Considerations for this Activity

During this experiment, we will be using a scale and ten pennies. There is no particular safety hazard associated with this activity. However, if someone has an allergy to copper or if there are small children in the household, one may wish to observe cautions to prevent an allergic reaction or choking hazard.

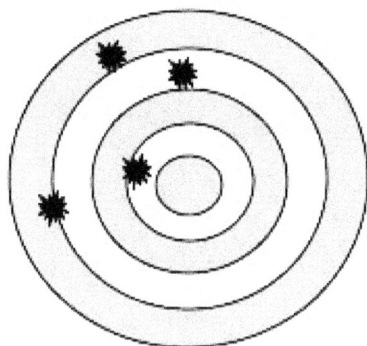

Not Accurate Not Precise

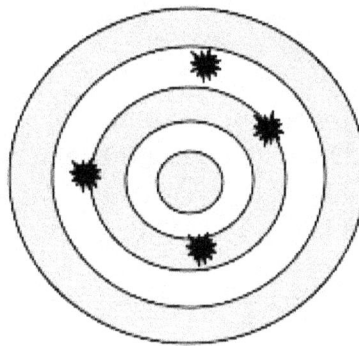

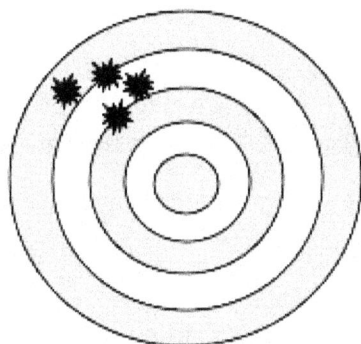

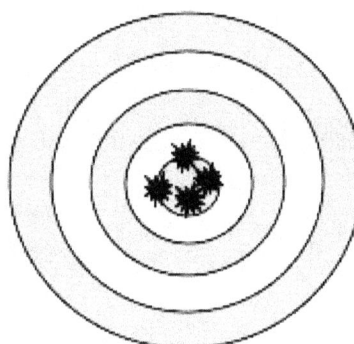

Accurate Not Precise

Not Accurate Precise

Accurate Precise

Visual comparison of accuracy and precision

Equipment required for the Activity:

- Ten post-1983 pennies (note this the United States Mint has different specifications for pre- and post- 1983 pennies).
- A scale that can read to 0.01 of a gram.
- Worksheet
- Calculator

Objectives:

- Determine Percent Error
- Determine Percent Difference
- Identification of sources of error
- Identification of bias

Procedure:

1. Collect ten post-1983 pennies.

2. Turn-on and tare (set your scale to zero) your scale.

3. Measure the mass of each individual penny and record the value in the data table on the worksheet.

4. Place all ten pennies on the scale and record the mass.

NOTE: The sum of the individual readings and the total mass reading may not agree. This is because each individual measurement introduces an error associated with the measurement. Thus, you will be comparing two different measurement techniques.

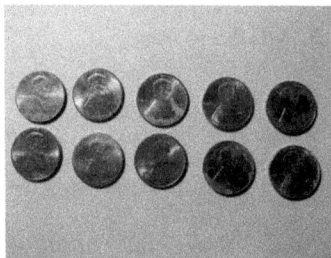

Above - Ten post-1983 pennies.

Right - Tared scale

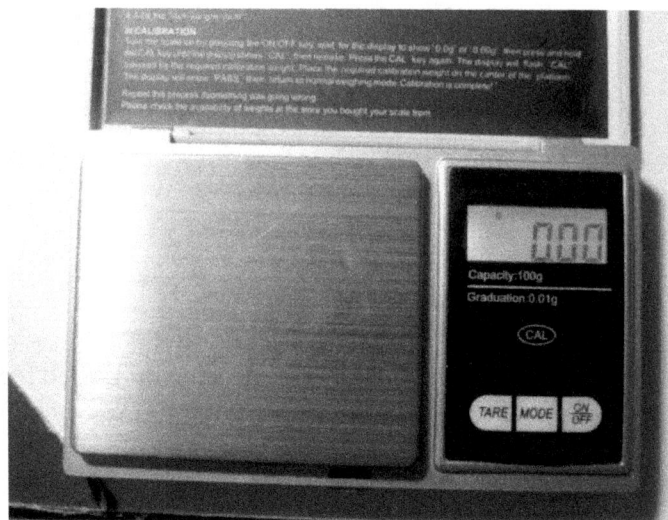

Procedure (Continued):

5. Using the individual measurements, calculate the total mass. This value will be designated CM.

6. Using the CM, calculate the average mass of a single penny – CM/10. This value will be designated ACM.

7. Using the measured value for the total mass of ten pennies, calculate the average mass of a single penny. This value will be designated AMM.

8. Calculate the percent difference – between the two ten penny mass measurements. See diagram.

9. Calculate the percent difference – between the two average penny masses.

Think About It -

Are these values the same? Do you expect them to be?
What is the percent difference telling you?
Do your numbers agree? Why or Why not?

10. Obtain the actual value of the mass of a penny from your instructor. Calculate the percent error, i.e. how your measured values compare with the true value.

Think About It -

Which value was more correct, i.e. had the lower percent error?
Do your values indicate that there is a bias?
If so, what is the potential source of bias?

Weighing a single penny (left) and ten pennies (right)

Accuracy, Precision and Distribution
Data Collection Sheet

Using only post 1983 pennies, weigh 10 pennies together and individually; complete the following data collection sheet.

	Post - 1983 Pennies
Coin	Mass (g)
1	
2	
3	
4	
5	
6	
7	
8	
9	
10	
Combined Mass as Measured of the 10 Pennies	

Using the data from the individual penny measurements, total the mass of the 10 pennies:

Total Mass of Pennies _____ CM

Calculate the average weight of a single penny using your total from above:

Average Weight of a Penny _____ ACM

Calculate the average weight of a single penny using the measured weight of the 10 pennies combined:

Average Weight of a Penny _____ AMM

Calculate the % Difference between your measured mass of 10 pennies and the calculated mass of 10 pennies (MM is measured mass)

$$\% \, Difference \; = \; \frac{|\, CM \, - \, MM \,|}{(\frac{|CM \, + \, MM|}{2})} \; x \; 100\%$$

% Difference = _____

Calculate the % Difference between the ACM and AMM:

$$\% \, Difference = \; \frac{|ACM \, - \, AMM \,|}{(\frac{|ACM \, + \, AMM|}{2})} \; x \; 100\%$$

% Difference = _____

Write down the actual mass of a penny as specified by the US Mint: _____ RM

Calculate the % Error between the AMM and the RM and the % Error between the ACM and the RM using the formula:

$$\% \, Error = \; \frac{|\, RM \, - \, Measured \, Value \,|}{(\, RM \,)} \; x \; 100\%$$

% Error = _____ (RM / ACM)

% Error = _____ (RM / AMM)

Summing It Up:

- Accuracy – how close is the measured value to the true value. This is evaluated using percent error. Accuracy is related to error.

- Precision – how close are the measured values to each other. You can see the precision in the variability of the individual measurements. The percent difference gives you an indication of precision between different types of measurements.

- Bias – is the measured result skewed from the true result. In the case of the penny experiment bias may appear due to oxidation or scratches over a set of coins.

Throughout this course, you will be using the concepts of accuracy and precision. You will be calculating percent difference and percent error depending upon the nature of the experiment. Thus, it is important to understand these concepts.

Laboratory #4
Understanding Measurements
Taking Measurements

Introduction

In simplified terms, the scientific process is a cycle of observations/tests, the development of a hypothesis to explain the observations, test again, reformulate the hypothesis, and repeat. Thus, the process requires three categories of skills – observation, communication and measurement. So, far in this course, you have been   exposed to how companies communicate information about products and how to find other sources of information. You have started your laboratory notebook, which will be used to record your observations. And, through your other educational activities, you have probably worked on other observation and communication skills.

During this course, you will be refining these skills as part of the investigation of the scientific process. However, this activity focuses on that other category, measurement. From a scientific perspective, measurement is a method of determining a quantity, capacity or dimension. There are a variety of systems used to determine a measurement. But, generally, a quantity, capacity or dimension is determined by using some type of reference. The reference or standard is something that has meaning and can be communicated with others.

For example, you have a box that needs to be mailed. How big is it? You could say that the box is bigger than a grapefruit, or you could provide the dimensions, height, width, and length in terms of inches. Inches are used in United States Customary System or English System. You could provide the dimensions in terms of centimeters which is part of the International System. (One inch is equal to 2.54 centimeters, exactly.) In our example, the box is compared to a standard that is familiar to the person that you are talking with, the grapefruit, inches, or centimeters.

Think about our box example. If you go to the post office, which provides a more accurate picture of the box: the box is slightly larger than a grapefruit or the box is 4 inches by 4 inches by 4 inches;? The answer is the description provided in inches. This measurement is considered to be more precise. It is better defined. There is less error, i.e. you have a better feel for exactly how big the box is because the standard of the grapefruit is variable.

Precision is critical in science. Precision is also communicated through the use of significant figures. A number or measurement should be written to express the degree of accuracy. Let's look at the following figure.

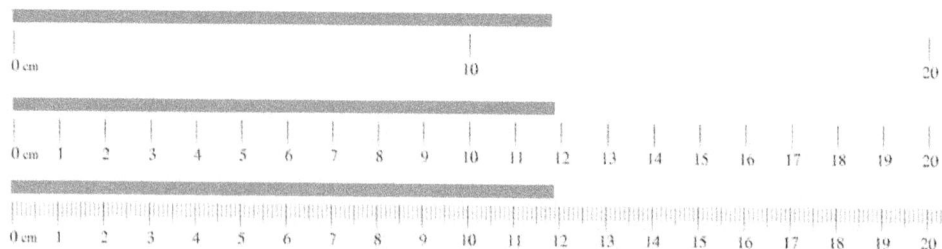

In this figure we have three rulers and we are measuring the bar. Our rulers are in centimeters, but come from different manufacturers. The first manufacturer is mostly concerned with larger measurements so only marks the ruler in ten centimeter increments. The second manufacturer is only concerned about moderate distances so only marks the rulers in centimeter increments. And, the third manufacturer makes typical desk ruler and provides millimeter markings. Measuring our bar with the three rulers we get the following three measurements:

12   centimeters (cm)
11.8   cm
11.85   cm

Before considering the precision of the numbers, let's discuss what is a significant figure or digit. In our first number there are 2 digits, in the second there are 3, and in the third there are 4. In these measurements they are considered significant because they have meaning and convey something to the reader. The last digit tells the reader where the error is. And, in the case of our ruler measurement tells us which digit had to be estimated.

When measuring, you can "eyeball" or estimate the number between the markings. So, in the first measurement the bar was greater than 10 cm but less than 20 cm and looked to be around where the 2 would be if we divided the space into ten. Similarly, in the second measurement, the end of the bar is greater than 11 cm but not quite to the 12 cm mark. Doing the same estimation, dividing the space into tenths, it looks to be about 8; so the measurement is 11.8 cm. Where the 8 becomes the estimated digit, and is where the error is located. By the time we use the more precise ruler, where the millimeters are recorded, the measurement has more significant digits because our reference is better. And, the measurement now can be written to another significant digit.

When measuring liquids, you will discover that the liquid does not form a flat surface. When confined in a cylinder, the liquid forms a meniscus (see the picture below) and generally curves down. (There are some liquids that curve up, but these are few.) You need to read the volume (because the graduated cylinder measures volume) at the bottom of the meniscus. The same "rule of thumb" as that with the ruler applies here as well. You need to read between the lines, and provide one more significant digit or figure than the scale of your cylinder.

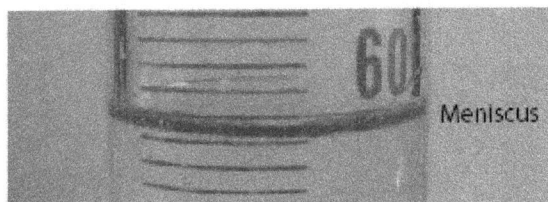

In this activity, you will be conducting a number of measurements, practicing a few laboratory skills, and doing some calculations with your measurements. So, let's get started.

The Rules
for
Working with Significant Figures

Rule 1 – *All non-zero digits are significant.*

Rule 2 – *Zeroes between non-zero digits are significant.*

Rule 3 – *Exact numbers have an infinite number of significant figures.*

Rule 4 – *When adding/subtracting position of the least significant digit determines the position of the result.*

Rule 5 – *When multiplying/dividing the result is reported has the same number of significant figures as the lowest number of significant figures used in the calculation.*

Working with Significant Figures

Counting Significant Figures:

When determining the number of significant figures, you must distinguish between digits that have meaning or zeroes that are place holders. Non-zero numbers are always significant (Rule 1) and zeroes between non-zero digits are always significant (Rule 2). Using scientific notation, eliminates confusion.

Examples:

0.0035	has 2 significant figures, the zeroes are place holders.
0.1230	has 4 significant figures, the zero in the ones location is a place holder, the zero at the end is significant.
12,300	has 3 significant figures, the zeroes are place holders.
504	has 3 significant figures.
5,430.	has 4 significant figures, the decimal signals that the zero is significant.
5.6×10^3	has 2 significant figures, all digits written in scientific notation are significant.

Note: if a number is EXACT such as a constant, there are an infinite number of significant figures. (Rule 3)

Working with Significant Figures

Calculating with Significant Figures:

Adding/Subtracting

Significant figures determine the precision of a number. When adding and subtracting, use the last digit retained is determined by the POSITION of the FIRST doubtful digit. (Rule 4)

Examples:

32.4 + 36.78 = 69.18   is reported as 69.2 as the 4 in 32.4 determines the position of the result.

16.5406 – 5.21 = 11.3306   is reported as 11.33 as the 1 in 5.21 determines the position of the result.

Multiplication/Division

Significant figures of the calculation can contain no more significant figures than the least number of significant figures used in the operation. (Rule 5)

Examples:

5.2 x 3.618 x 91.8 = 1727.088 (on calculator)
should be reported as 1,800 as 5.2 has only 2 significant figures.

(6.314 x 98.651)/15.8831 = 39.21667773 (on calculator)
should be reported as 39.22 as 6.314 has the least number
of significant figures at 4

General Safety Considerations for this Activity

During this laboratory, you will be using a ruler, scale, and graduated cylinder. You will be measuring some items located in your home. While there is no particular hazard associated with this activity, it is time to start practicing some good laboratory safety habits, like wearing gloves and safety goggles. Also, you should complete a hazard analysis in your notebook before you begin.

Equipment required for the Activity:

- A scale that can read to 0.01 of a gram.
- Graduated cylinder
- Ruler marked in centimeters
- A regular piece of copy paper (8 ½ x 11 inches)
- A one-cup measuring cup – with both ounces and milliliters (ml)
- Calculator
- Water (tap water can be used for this activity)
- A small block of wood (you can use a pad of sticky notes or some other block like object but it needs to be less than about 20 grams)
- Nail or screw
- Five post-1983 pennies
- Laboratory notebook

Objectives:

- Practice measuring dimensions, volume and mass
- Calculate volume and density based on measurements
- Compare volume determinations
- Develop a frame of reference between English and International measurement systems

Recording information in your laboratory notebook.

Should you write down a number or some other piece of information incorrectly in your notebook, the proper procedure for correcting it is to draw a line through the incorrect information, initial it and write the correct information beside it.

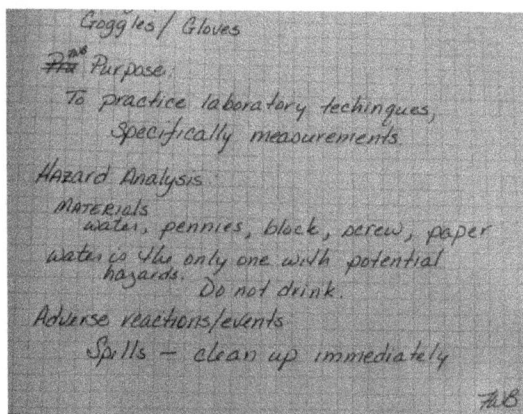

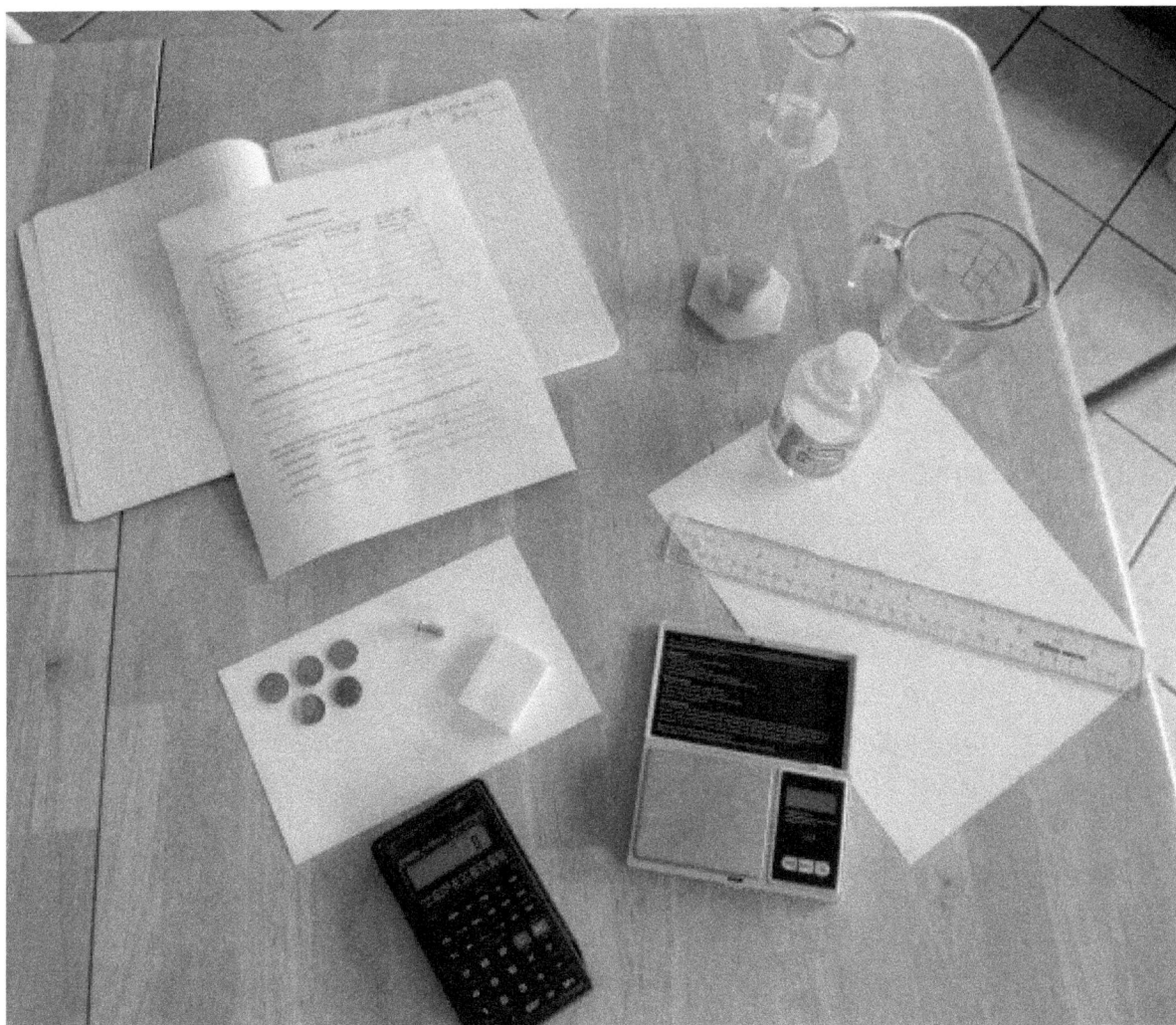

Activity

Measuring Weight/Mass

1. Gather the materials that you are going to weigh – block, nail or screw, and the five post-1983 pennies.

2. Review the instructions that have come with your scale.

3. Weigh each item and record the weight in your laboratory notebook.

NOTE: You need to record all of the digits displayed on your scale.

Think About It -

What is the difference between mass and weight?

Measuring Length

In this section you are going to measure length in centimeters (cm) and compare cm to inches.

1. Using the ruler, measure the length and width of paper in cm, being sure to use the appropriate number of significant figures. Record these measurements in your laboratory notebook.

2. Using the ruler, measure the length, width and height of the block. Record these measurements in your notebook.

Think About It -

How many centimeters equaled the 8 ½ inches, width of the paper? How many centimeters equaled the 11 inches, length of the paper? Can you determine a conversion factor for inches to centimeters using your data?

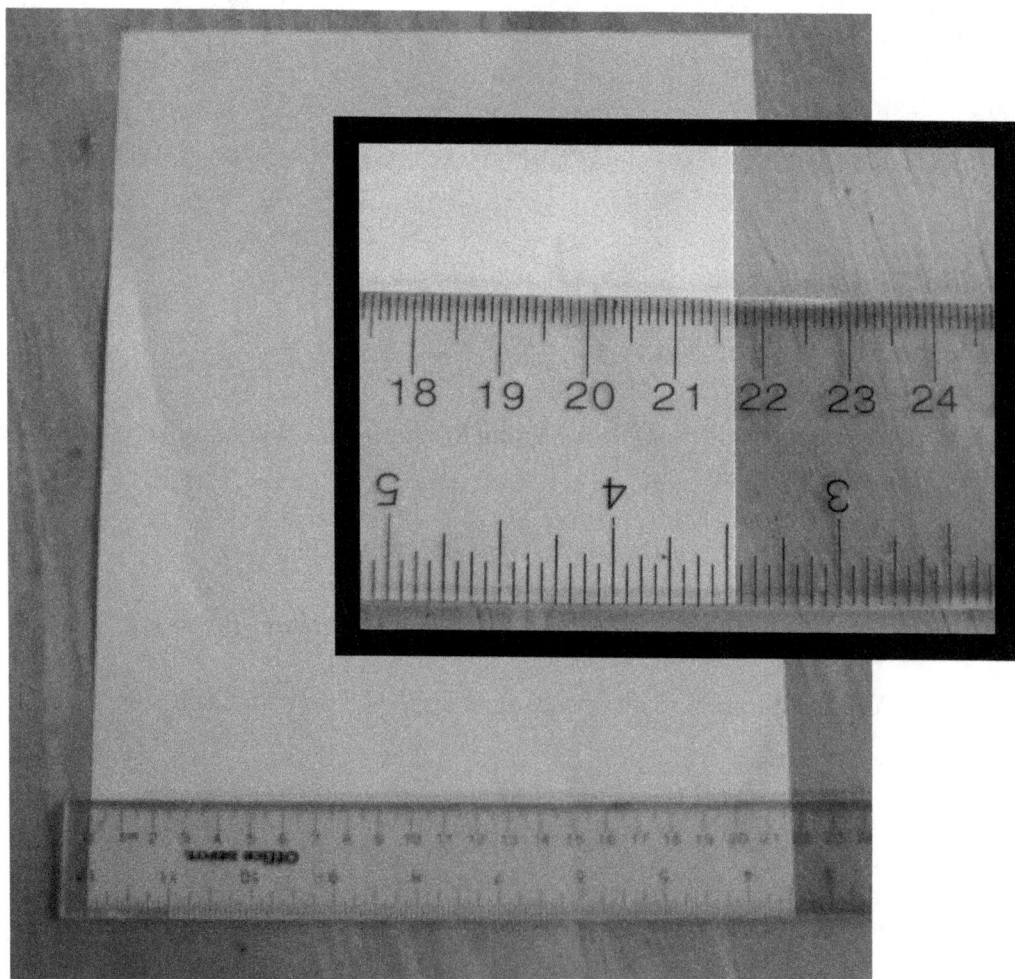

Example of notes in the laboratory notebook.

Measuring Volume

Numerous types of volumetric measuring devices or apparatus can be used to determine a volume of a liquid. In the kitchen, volume is typically measured in ounces. When measuring using the System International, volume is measured in milliliters. On most measuring cups in the kitchen today, you can measure in either ounces or milliliters. The abbreviation for milliliters is mL or ml. In the laboratory, scientists usually measure a liquid using a graduated cylinder.

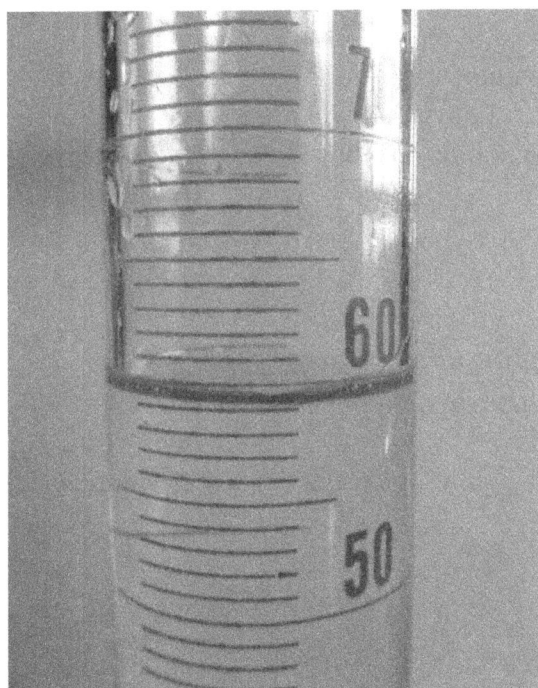

How to read a graduated cylinder

Just like in the example of the ruler, the cylinder is marked in increments. In the case of the graduated cylinder, these increments are in ml. And, just like in our example with the ruler; you will read the measurement to the appropriate significant figure. In the case of a 100 ml graduated cylinder, you will be able to measure to the 0.1 digit or estimate the level between the individual ml. (See Figure)

Volume can also be calculated by using a linear measurement. For example: your block has height, width and length. The volume of a block is determined by multiplying the height times the width times the length. You can determine the volume of many objects, just by knowing the dimensions and looking up the formula for that particular object. Objects that the volume can be easily calculated include: cubes, blocks, spheres, cones, and cylinders. If the volume is measured in centimeters (cm), you can easily convert between the linear type measurement and the liquid measure in ml. One cubic cm or cm3 of water is defined as 1 ml. NOTE: the units of volume for solid objects in linear measures such as meters or centimeters will be represented as a cube of the length. This is because length (cm) x width (cm) x height (cm) = volume (cm3). Knowing that the volume is a cube of the length measurement will help you in your unit analysis and will allow you to check your work.

You can measure non-standard shapes such as screw, by using the displacement method. In this method, you will fill your graduated cylinder with enough water to cover the screw. (This is initially done without the screw being placed in the cylinder.) Record the volume in the graduated cylinder. Then carefully drop the screw into the liquid. (You don't want to lose any of the water from the cylinder. If you do lose water, your result will be inaccurate.) Record the new volume measurement. This measurement includes the volume of the water plus the volume of the screw. By subtracting the volume of the water that was initially in the graduated cylinder, you obtain the volume of the screw.

In this portion of the laboratory, you are going to determine the volume of a liquid, your block, the nail or screw and the five post-1993 pennies.

Volume of a Liquid

1. Obtain ¼ cup of water in your measuring cup.

2. Pour the water into the graduated cylinder and record the volume to the proper significant figure in your laboratory notebook.

> *Think about it –*
>
> *How did your reading on the graduated cylinder compare with the ml markings on the measuring cup.*

3. Leave the water in the graduated cylinder. You will be using this to determine the volume of the five pennies.

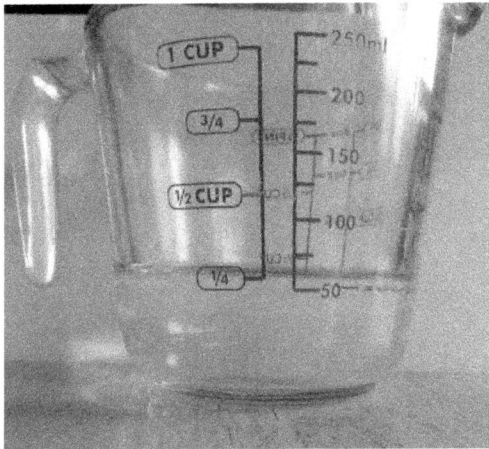

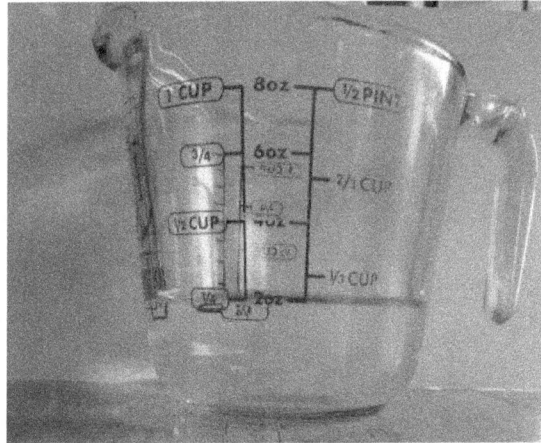

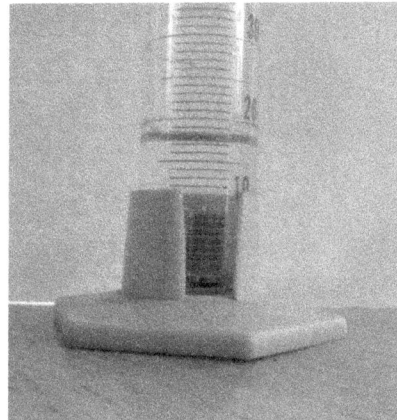

MEASUREMENTS LAB (cont)   DATE

Volume

Using measuring cup 1/4 cup ≅ 65 ml
            w/WATER

Graduated cylinder   =   59.0 ml

   Volume initial 59.0 ml
   Volume final (w pennies)   60.9 ml

      Volume of pennies   60.9 - 59.0 = 1.9 ml

Screw
      Volume initial = 17.5 ml
      Volume final   = 18.0 ml

      Volume of screw = 0.5 ml

Discussion/Observations:

Volume of the Block

Using your length data for the block, calculate and record the volume of the block. (Volume = length x width x height)

Volume of the Pennies and Either the Nail or Screw

1. Verify the volume of the water in the graduated cylinder. This is your initial volume for the determination of the volume of the pennies.

2. Carefully drop the five pennies into the cylinder with the water. Record the volume. This is the final volume.

3. Determine the volume of the five pennies by subtracting the initial volume from the final volume. Document the calculation and result in your laboratory notebook.

4. Empty the graduated cylinder of the water and pennies, being careful not to lose the pennies down the drain.

5. Fill the graduated cylinder to a level that will cover the nail or screw that you are measuring the volume.

6. Record the initial volume.

7. Carefully drop the nail or screw into the cylinder. Record the final volume.

8. Determine the volume by subtracting the initial volume from the final volume. Document the calculation and result in your laboratory notebook.

9. Empty the graduated cylinder of the water and the nail or screw. Again being careful not to lose the nail or screw down the drain.

Final Considerations

1. Once you have completed all of the activity, verify that you have recorded all of the data in your laboratory notebook correctly. If you need to repeat a measurement do so and record it properly.

2. In your discussion section, document:

> * Any observations that you made.
> * Any conversion factors derived.
> * Comparisons between the obtained conversion factors and
>   reference conversion factors.
> * Any potential sources of error.

3. Once complete, clean up your area and put away all the materials. Properly dispose of any waste products. For this activity, water is the only waste material. You can dispose of this into a sink.

Summing It Up:

In conducting this activity, you practiced a number of laboratory techniques to measure volume and mass as well as length. You began to record data in your laboratory notebook, and documented calculations.

As part of the laboratory, you were introduced to significant figures. The number of the digits presented in the data represent the level of accuracy of the measurement. This has introduced the concepts of precision and error. Additionally, you have obtained some experience in how to handle significant figures in calculations.

You will be using the information that you gathered in this laboratory in another laboratory a bit later on in the course.

Laboratory #5
Exploration of Physical Properties
Density

__Introduction__

Matter is anything that occupies space and has mass. Matter can be in one of four different physical states: plasma, gas, liquid or solid. (In general, most materials will be in one of three physical states, gas, liquid or solid. In this course, we will focus on the three common physical states.) In a different experiment, you will investigate the states of matter. But, in this experiment you will be introduced to one physical property of matter – density.

Matter can be described as either being comprised of a pure substance or a mixture of substances. A pure substance is one that all of the material is comprised of a single type of atom or molecule. A pure substance can be water with no contaminates, a block of aluminum, or the helium gas in a balloon. Mixtures are comprised of more than one substance such as sea water, bronze or air.

Pure substances can either be elements or compounds. For example, table salt is made up of sodium (Na) and chlorine (Cl) atoms. Salt is a compound. So while, a salt shaker may only contain a pure substance, it does not consist solely of a pure element.

Elements, i.e. the atoms on the periodic table, can be described by a number of physical properties. Compounds such as table salt can also be described by its physical properties. There are a number of physical properties and include the following: odor, color, melting point, boiling point, and density. We can use these physical properties to help us identify unknown materials. During this laboratory, you are going to be introduced to a number of physical properties, and compare your observations with established standards.

General Safety Considerations for this Activity

In this laboratory, you will be using general household materials, thus, there is no particular safety risk associated with this activity. However, as you have seen previously, it is a good idea to be use proper laboratory technique and use the appropriate personal protective equipment. For this activity, you should wear gloves and safety goggles.

When completing the hazard analysis for this laboratory, think about the materials that you will be using: water, table salt, and baking soda. Are there any hazards associated with them?

Equipment required for the Activity:

- A scale that can read to 0.01 of a gram.
- Graduated cylinders (100 mL and 10 mL)
- 5 pre-1983 pennies
- 5 post-1983 pennies
- Table salt
- Baking Soda
- Water*
- Your laboratory notebook
- Internet access

* For this laboratory, it is suggested that distilled water be used. Tap water or other water may have contaminants that may skew the laboratory results.

Objectives:

- Initial discussion of properties of matter
- Understanding of density
- Understanding of the difference between accuracy and precision
- Determination of pure versus mixtures

Procedure:

1. Read and review the entire activity before proceeding. This will allow you to complete your hazard analysis.

2. Using your computer, obtain a SDS for water, table salt, and baking soda.

3. Review the safety hazards, and complete the hazard analysis in your laboratory note book. Do not forget to include the pennies in your hazard analysis. Pennies can be a choking hazard.

4. From the SDS, document the following information in your laboratory notebook for water, table salt, and baking soda:

Color/Appearance: _____

Physical State: _____

Density: _____

Molecular Weight: _____

Formula: _____

This information will be used later in this laboratory. Some things to be aware of when documenting this information. Be sure to include the reference.

Also, be sure to include any units provided. For example: your density on the SDS may be in lb/ft^3 or in g/cc (grams/cm^3). If your density is in lb/ft3 or some other unit – please convert to grams/cm^3 or grams/mL. See inset for conversion help.

Your table in your laboratory notebook, might look like this:

Table Salt

Information from Morton Salt SDS

Color/Appearance: _____

Physical State: _____

Density: _____

Molecular Weight: _____

Formula: _____

Procedure (Continued):

5. Gather your materials to conduct the laboratory.

6. For this part of the laboratory, use the 10 mL graduated cylinder. Verify that your graduated cylinder is dry. Weight the graduated cylinder and record the weight. You will be using this weight for several calculations.

7. Measure out 1 level teaspoon of salt. Record the following observations for the salt:

      Color/Appearance

      Physical State

8. Pour the teaspoon of salt into the graduating cylinder. GENTLY, tap the graduated cylinder so that the salt compacts. When the salt has compacted and the top layer is level, weigh the graduated cylinder with the salt. Record this weight in your notebook. (Remember your significant figures.)

9. Read the volume of the salt by reading the graduated cylinder.

10. Calculate the mass of the salt. (The mass of the graduated cylinder plus the salt minus the mass of the graduated cylinder empty.)

11. Calculate the density of the salt. Density is the property of matter indicating the mass per unit volume. For this activity, density will be in units of grams/mL. Density is calculated by taking the mass of the material divided by its volume. Record this value in your laboratory notebook.

$$Density = \frac{mass}{volume}$$

12. Discard the salt into the trash.

13. Repeat Steps 7 through 12 for the baking soda.

14. Rinse the graduated cylinder with water (3 times) after completing the baking soda measurement. Dry the cylinder.

15. Measure 1 teaspoon of water. (For this laboratory, it is suggested that distilled water be used. Tap water or other water may have contaminants that may skew the laboratory results.)

Procedure (Continued):

16. Pour the 1 teaspoon of water into the graduated cylinder. Read the level of water in the graduated cylinder. Record this measurement in your laboratory notebook.

17. Record the same observations that you did for the salt and the baking soda for the water. (Step 7)

18. Weigh the graduated cylinder with the water. Record the weight.

19. Determine the mass of the water. You will use the same method as in Step 10.

20. Determine the density of the water, using the same method as in Step 11.

21. Discard the water into the sink.

22. Determine the densities of a single pre-1983 and a post-1983 penny by using the mass of five pennies (both pre- and post-) and determining the volume using water and a graduated cylinder. Document your procedure in your laboratory notebook. Show your calculations.

23. Clean-up your work area and prepare to do some further analysis.

24. Calculate the densities for water, salt, and baking soda using the teaspoon measure after converting teaspoons into mL. The conversion factor for teaspoons to milliliters is 1 teaspoon is equal to 4.92892 milliliters. Record this measure for density.

25. Calculate the % Difference between the two methods you used to determine the density. The two methods are 1) using the graduated cylinder to determine volume and 2) using the teaspoon to determine volume.

26. Using the information from the SDS for the density of the three materials, calculate the % Error between the observed densities for the three substances from Steps 7 through 20.

$$\% \; Difference = \frac{|Value \; 1 - Value \; 2|}{(\frac{|Value \; 1 + Value \; 2|}{2})} \; x \; 100\%$$

$$\% \; Error = \frac{|Real \; Value - Measured \; Value|}{Real \; Value} \; x \; 100\%$$

Analysis of the Data from this Activity:

1. How did the densities compare between the graduated cylinder method and the teaspoon method? Which of these is the more accurate measurement? (Hint: Which has more significant figures?)

2. How did the densities compare to the "known" densities provided in the SDS? Why were there differences? (Hint: What sources of error may be present?)

3. Compare the densities of the pre-1983 and post-1983 pennies. Are they the same?

4. Using the post-1983 penny density, compare your determined density with the density obtained using the U.S. Mint specifications. Show your % Error.

5. How does the density of the penny compare to the density of copper? (Copper has a density of 8.94 g/mL at 25 degrees C.)

6. Can you determine the percentage of copper in a post-1983 penny?

References
US Mint Coin Specifications:
http://www.usmint.gov/about_the_mint/?action=coin_specifications

Laboratory #6
More Physical Properties
And
States of Matter

Introduction

In our last laboratory, there was a brief discussion about the states of matter. In that discussion, you were introduced to the three common states of matter: solid, liquid and gas. In this laboratory, you will investigate these three states of matter by looking at two additional physical properties of compounds. The two physical properties to be investigated are freezing point (melting point) and boiling point. These physical properties indicate when a material makes a transition between two states of matter. For the freezing point/melting point, the transition is between a solid and a liquid. For the boiling point, the transition is between a liquid and a gas.

You are probably already familiar with the terms solid, liquid and gas. But, let's look at these terms as they are used in chemistry. For the physical state of a material to be defined as a solid, it must be firm and stable in shape. A liquid takes the shape of its container. It flows but has a constant volume. A gas can expand to fill the entirety of its container, and is less dense than either a solid or liquid. Let's look at the three states of matter for water:

Solid   Ice
Liquid   Water
Gas   Steam

The states of matter can also be visualized.

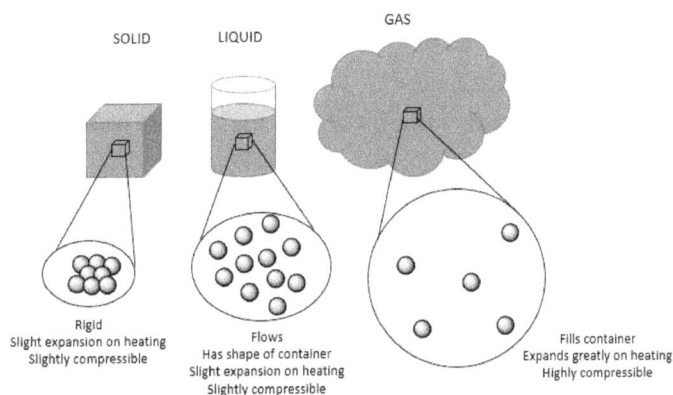

SOLID — LIQUID — GAS

Rigid
Slight expansion on heating
Slightly compressible

Flows
Has shape of container
Slight expansion on heating
Slightly compressible

Fills container
Expands greatly on heating
Highly compressible

Freezing Point & Supercooling

The freezing point is defined as the point at which a liquid turns to solid, and in theory the melting point should be the same. In this laboratory, and to the precision of our measurement this is the case. However, with better equipment, a slight difference can be observed. Some liquids can be "supercooled" which means that they can remain a liquid without freezing. You may want to investigate the phenomena of supercooling on the Internet.

General Safety Considerations for this Activity

During this laboratory, you will be using a heat source (your stove or hot plate). Thus, caution will be required to ensure that you do not get burned. Additionally, as paraffin wax melts; it becomes flammable. So additional caution needs to be taken to ensure that you do not have a fire. As with your previous laboratories, you will need to complete your hazard analysis and where your gloves and safety goggles.

Equipment required for the Activity:

- Your laboratory notebook
- Water (distilled water and tap)
- Table salt (approximately 2 tablespoons) to make the ice bath
- Ice
- Paraffin Wax (Approximately 1 oz) (Canning wax)
- 2 disposable tart tins (or test tubes)
- A pot that can be used on the stove
- Digital thermometer that can read in degrees C
- Stove or hot plate
- Stirring spoon
- Hot pad or tongs to remove the tart tin from the hot water.

Objectives:

- Measure physical properties
- Use an ice and water bath

Procedure:

1. Read through the entire laboratory prior to starting.

2. Using the Internet, obtain an SDS for Paraffin Wax. From the SDS, document the following information in your laboratory notebook for paraffin wax:

Color/Appearance: _____

Physical State: _____

Density: _____

Molecular Weight: _____

Formula: _____

Boiling Point: _____

Flash Point: _____

Melting Point: _____

Did you notice something about the formula of paraffin wax? Paraffin wax, while it has a CAS number for identification, is not a single compound. It is a mixture of hydrocarbons. The mixture contains a number of different molecules in a specific range defined by the number of carbons and the boiling point.

Hydrocarbons

Hydrocarbon is a term used in organic chemistry refering to compounds made entirely of carbon and hydrogen. These materials are generally used as fuels.

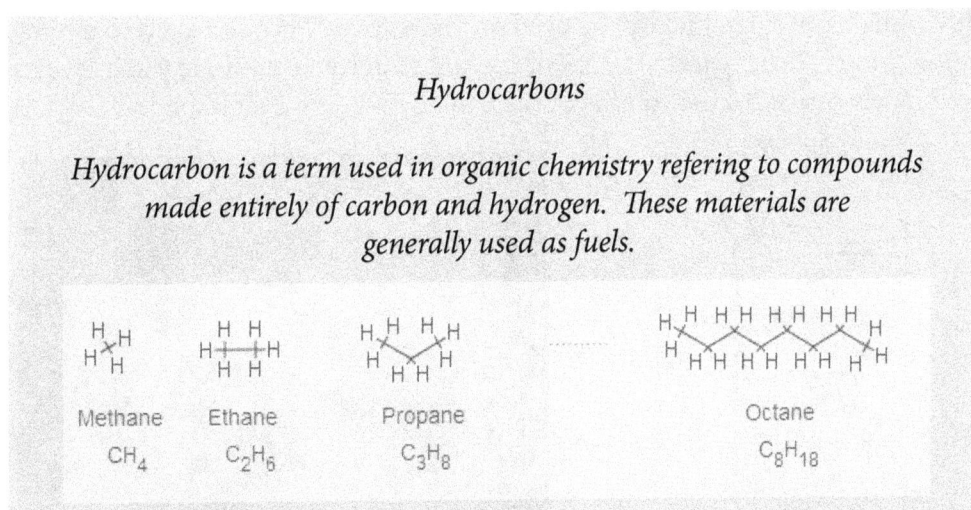

Methane CH_4 Ethane C_2H_6 Propane C_3H_8 Octane C_8H_{18}

Procedure (Continued):

3. Refer back to the water SDS that you obtained in a previous laboratory, record the following in your laboratory notebook:

> Boiling Point: _____
> Flash Point: _____
> Melting Point: _____

Flash Point

The flash point of a volatile (the tendency of a substance to vaporize) material is the lowest temperature at which it can vaporize to form an ignitable mixture in air, i.e. the point where the vapors can catch fire if exposed to an ignition source like a flame.

Using the definition of flash point, would you expect there to be a flash point for water? Did your SDS list the melting or freezing point for water? Make sure you document any changes in your laboratory notebook.

4. Complete your hazard analysis.

5. Prepare for the laboratory. Gather all the equipment that you will need using your equipment list. You can place your paraffin wax into one of the tart tins.

6. Building the ice bath: In a laboratory, an ice or water bath is built using a large beaker. (See Figure.) For this laboratory, you are going to build an ice bath/water bath in a small sauce pan that can be used on your stove. Choose a sauce pan that will allow you to "float" your tart tin. Fill the pan with water until the water level is approximately one-third the volume of the pan and add the 2 tablespoon of salt. Stir until the salt dissolves.

You will investigate the effects of salt in water on the freezing point and boiling point of water in another laboratory.

Once the salt is dissolved, add enough ice to raise the level of the water to approximately halfway in your pan. Stir to facilitate the temperature equilibration. You want the water and ice to equilibrate in temperature. Once the temperature has stabilized you can begin the rest of the laboratory.

Equilibrium

Equilibrium is used to describe a state, such as temperature, of a system (in this laboratory, the system would be the pan, water, and ice) when all parts are the same. Thus, a system is said to be in equilibrium if all parts of the system are at the same temperature.

7. Record the starting temperature of your water/ice bath in your laboratory notebook.

Think about It:

What did you observe about the temperature of the water/ice bath? Is this something you expected? Why or why not? Record this information in your laboratory notebook.

8. Place the water/ice bath on to the stove.

9. Place a couple of ice cubes into the unused tart tin. And, float the tart tin with the ice in the water/ice bath.

Procedure (Continued):

10. Turn on the stove and begin heating the water/ice bath. You will want to use a low heat to heat the water/ice bath. If you heat the water/ice too fast you are likely to introduce error as you will not be able to accurately measure the temperature when you observe the change.

11. Observe the ice cubes in the tart tin. Record the temperature in your laboratory notebook of the water/ice bath when the ice begins to melt.

12. Shut off the heat to the water/ice bath, while you prepare for the next two steps of the laboratory. You can leave the pan on the stove during this period.

13. Remove the tart tin with the ice from the pan. Discard the ice into a sink. Rinse the tart tin with distilled water. Once rinsed, add approximate ¼ cup of distilled water to the tin and set aside.

14. Place the tart tin with the paraffin wax into the water/ice bath. Turn on the stove and begin heating the water/ice bath as you did in Step 10.

15. Observe the paraffin wax. Record the temperature in your laboratory notebook of the water bath (the ice will have completely melted by now) when the paraffin wax begins to melt.

CAREFUL: AT THE POINT THAT THE PARAFFIN WAX BEGINS TO MELT THE WATER WILL BE HOT.

16. Exchange the tin with the paraffin wax, with that of the distilled water. CAUTION THE WATER WILL BE HOT. Set the paraffin wax tin aside and out of the way while you continue the laboratory.

CAUTION THE WATER WILL BE HOT FOR THE REMAINDER OF THE LABORATORY. BE AWARE OF THE POTENTIAL FOR BURNS.

17. Continuing to heat the water bath, observe the distilled water in the tart tin. Record the temperature in your laboratory notebook of the water bath when the distilled water begins to boil. SHUT OFF THE HEAT TO THE PAN.

Think about It: What did you observe about the water bath at the point when the distilled water began to boil? Did you expect this? Why or Why not? Record this information in your laboratory notebook.

18. Allow the pan to cool. While cooling, clean-up and put away the other laboratory equipment. You can discard the tart tin with the paraffin wax into the trash. Once the pan has cooled, you can complete your clean-up.

Reviewing and Analyzing the Data:

In Steps 11 and 15, you observed the melting point of water (Step 11) and the melting point of paraffin wax (Step 15). How did your measurements compare with the established values that you obtained from the SDS? (Hint: you should calculate the percent error as part of your response.)

In Step 17, you observed the boiling point of water. How did your measurement compare with the established value you obtained from the SDS? (The percent error should be included as part of your discussion.)

What sources of error might have been present in this laboratory?

What other observations did you make?

Summing it Up:

Physical properties are used to identify and characterize elements, compounds and mixtures. In this laboratory, you observed and measured melting point and boiling point. These are two additional physical properties of a compound or element. These physical properties indicate when a substance moved from one physical state to another. During this laboratory, you should have been able to observe three states of matter: solid, liquid and gas.

You were also introduced to another physical property, flash point. Flash point is a characteristic that is used to determine how flammable or combustible a material is. It is an important physical property and is a good indicator of potential safety hazards. Fire is the hazard of concern when dealing with flash point.

You were introduced to hydrocarbons. Hydrocarbons are used by many as a source of fuel and include both pure substances and mixtures. Natural gas, gasoline, diesel, and wax are all hydrocarbon mixtures. You were also introduced to how hydrocarbon mixtures may be classified, by a range of carbon number.

Hopefully, you were able to observe that salt impacted the melting point and boiling point of water. You will investigate this impact in a later laboratory.

Laboratory #7
Moles, Molecular Weight, and Molarity

Introduction

So far in this course, you have explored a number of topics related to finding information about elements and compounds (molecular and ionic). Additionally, you have investigated the concepts of accuracy and precision as well as begun to learn some valuable laboratory skills. During these investigations, your focus has been primarily observational as it relates to chemistry.

This means you have been focused on the physical properties like density and melting point or have made observations about the state of the material, e.g. is it a solid or a liquid. But, there is some additional information that you need to know before you can really begin to experiment in chemistry.

Chemistry like math or music has its own language and ways of presenting information. You have already seen this as you collected chemical information. Recall when you looked at the SDS for a particular chemical, its chemical formula was listed. The chemical formula is short hand for a listing of the elements and their ratios (a recipe if you will) to each other to form that specific compound. Additionally, you also were able to find the compound's molecular weight.

There is one other term that is used in chemistry known as molarity. Molarity refers to the concentration of a compound in a solvent. In your chemical scavenger hunt, you may have run across information on the label that said "contains 2.5% potassium from potassium chloride" or "contains 5% acetic acid in water." This is just another way of looking concentration, but for the chemist molarity makes it easier to work with solutions when performing chemical reactions.

In this laboratory, you are going to conduct several different activities to help you understand the mole and molecular weight. Additionally, you will be introduced to molarity, chemical equations and stoichiometry.

The Mole

In the previous activities, you have been introduced to elements. Elements are a pure chemical substance represented as atoms on the periodic table. The Periodic Table is arranged by atomic number, the number of protons in the nucleus of the atom. The elements in each column of the Periodic Table have similar properties and will react with other elements in a similar matter. Thus, the Periodic Table is an excellent reference for chemists. And, depending upon which version of the Periodic Table you have, there will be a great deal of information communicated in each box representing the element, more about this in just a minute.

While chemists try to understand what is happening at an atomic or molecular scale, it is very hard to work that way in the laboratory. Think about it – have you ever held a single atom? Or, a single molecule? And, until very recently, no one had ever been able to capture pictures of a single atom or molecule. (If you want to see pictures of atoms or molecules, you can do an Internet search.) So for a chemist to conduct work in the laboratory, there must be a way to relate atoms to a mass that can be readily measured.

This is where Amedeo Avogadro comes into the picture. Avogadro was an Italian physicist who by working with gases observed that the same volume of gas at the same temperature and pressure had a constant relationship with other gases with differing molecular weights (or mass). Or thinking about it differently, one liter of argon gas had the same number of atoms as one liter of helium gas, even though the mass of an atom of argon is heavier than the mass of an atom of helium. This observation led to the concept of a mole.

A mole is a unit of measure much like a dozen or a ton is a unit of measure. A mole is defined as the amount of any substance that contains as many particles (atoms, molecules, ions, electrons) as there are atoms in 12 grams of pure carbon-12, the isotope of carbon with relative atomic mass of exactly 12 by definition. The number of particles is named after Avogadro, and is called Avogadro's number. Avogadro's number is approximately 6.02214×10^{23}.

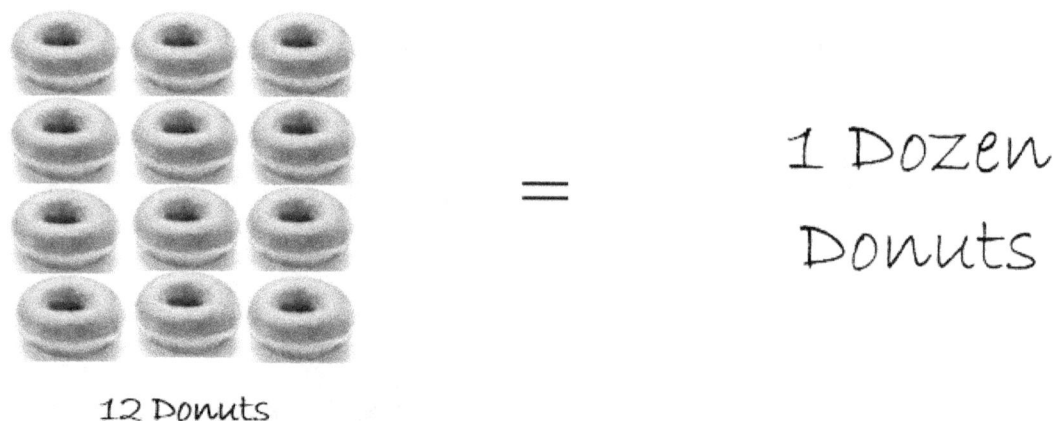

$=$ 1 Dozen Donuts

12 Donuts

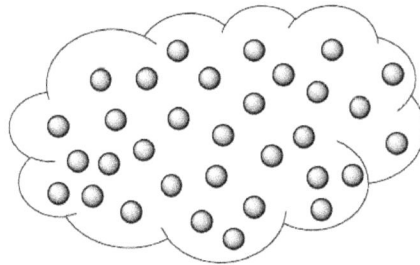

6.02214179 x10²³ of these ◯ = 1 mole of ◯

Where ◯ can be particles, atoms, or molecules.

A Mole

6.02214179 x10²³ is known as Avogadro's Number

The number is defined as the amount of substance (particles, molecules, or atoms) as there are atoms in exactly 12 grams of Carbon-12 atoms.

The units of Avogadro's number are
$$\frac{\# \text{ of particles, atoms or molecules}}{\text{mole}}$$

So you can have a mole of M&Ms, a mole of sodium chloride, or a mole of copper. Just like you can have a dozen eggs, or a gross of elephants. A mole is a means of counting particles. But, Avogadro showed that this number was related to the atomic mass or molecular weight and this information is recorded on the Periodic Table.

aluminum
13
Al
26.98

Name of the element

Atomic number which is equal to the number of protons in the atom

The symbol that represents the element

Atomic weight

Atomic weight is the number of grams per mole of that particular element. Or

$$\frac{\text{grams of Aluminum}}{\text{Mole}}$$

Using aluminum as an example, you can covert between moles, grams and number of atoms.

The atomic weight for aluminum is 26.98 grams/mole or

$$26.98 \frac{g}{mol}$$

Let's say you were given a pure aluminum token that weighed 100. grams. Determine the number of moles and the number of aluminum atoms in the token.

$$100. \text{ grams of aluminum } x \; \frac{1}{26.98 \frac{g}{mol}} = 3.71 \text{ moles of aluminum}$$

$$3.71 \text{ moles of aluminum } x \; 6.022 \; x \; 10^{23} \; \frac{\# \text{ of aluminum atoms}}{mole} = 2.23 \; x \; 10^{24} \text{ aluminum atoms}$$

So, the atomic mass unit which is used interchangeably with the term molecular weight, means that for sodium, one mole of sodium atoms has a mass of 22.99 grams, or $6.022 \; x \; 1023$ atoms of sodium has a mass of 22.99 grams. We can use the information from the period table to determine the mass of compounds, i.e. the molecular weight. (See Calculating Molecular Weight.)

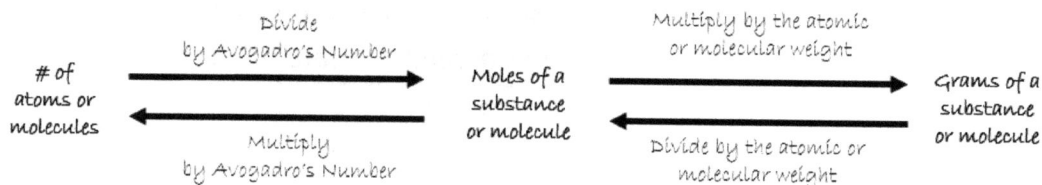

| # of atoms or molecules | Divide by Avogadro's Number → ← Multiply by Avogadro's Number | Moles of a substance or molecule | Multiply by the atomic or molecular weight → ← Divide by the atomic or molecular weight | Grams of a substance or molecule |

Isotope

If you think of an atom with a specific atomic number, for example hydrogen with an atomic number of 1; there will be one proton in the nucleus of the atom. However, the nucleus of an atom is made up of protons and neutrons. So, an isotope is an atom that has the same properties and number of protons but has a differing number of neutrons. For hydrogen, the common isotope is deuterium which has one proton and one neutron in its nucleus.

Calculating Formula/Molecular Weight

Once you have the correct chemical formula, then calculating the formula weight becomes a fairly simple process.

Step 1 – Determine the number of each atom type.

Step 2 – Using the Periodic Table, obtain the atomic mass (amu or grams/mole (g/mol)) for each atom type.

Step 3 – Multiply the atomic mass for the element by the number of atoms contained in the formula.

Step 4 – Sum up all of the masses to obtain the mass for the formula.

For example: Sucrose, table sugar, has the following molecular formula:

$$C_{12}H_{22}O_{11}$$

Which means it has 12 carbon atoms, 22 hydrogen atoms and 11 oxygen atoms in a single molecule of sucrose (Step 1). From the Periodic Table, the elements have the following atomic masses (g/mol) (Step 2):

Carbon	12.01 g/mol
Hydrogen	1.01 g/mol
Oxygen	16.00 g/mol

To calculate the molecular formula:

12 atoms of C x 12.01 g/mol	144.12 g/mol
22 atoms of H x 1.01 g/mol	22.22 g/mol
11 atoms of O x 16.00 g/mol	176.00 g/mol
Total	342.34 g/mol
	- the molecular weight for sucrose

Or, you can say that 1 mole of sucrose will have a mass of 342.34 grams or that 6.022×10^{23}
molecules of sucrose have a mass of 342.34 grams. (Recall that one pound is 453.59 grams, therefore 1 mole of sugar is approximately ¾ of a pound of sugar.)

Some additional examples:

Bleach
Molecular Formula
NaClO

Number of Atoms in Formula

Na	1 atom
Cl	1 atom
O	1 atom

sodium 11 Na 22.99   oxygen 8 O 16.00   chlorine 17 Cl 35.45

Atomic weight in either amu or g/mol

Molecular or Formula Weight =

Number of atoms of Na x atomic weight of Na	1×22.99 g/mol
+	
Number of atoms of O x atomic weight of O	1×16.00 g/mol
+	
Number of atoms of Cl x atomic weight of Cl	1×35.45 g/mol
Total molecular weight	74.44 g/mol

Washing Soda
Molecular Formula
Na_2CO_3

Number of Atoms in Formula

Na	2 atoms
C	1 atom
O	3 atoms

sodium 11 Na 22.99   oxygen 8 O 16.00   carbon 6 C 12.00

Atomic weight in either amu or g/mol

Molecular or Formula Weight =

Number of atoms of Na x atomic weight of Na	2×22.99 g/mol
+	
Number of atoms of O x atomic weight of O	3×16.00 g/mol
+	
Number of atoms of C x atomic weight of C	1×12.01 g/mol
Total molecular weight	105.99 g/mol

H_2O $\qquad$ $Ca(CO_3)_2$ $\qquad$ CH_3COOH $\qquad$ $(NH_4)_2SO_4$

2 H	1 Ca 2 CO$_3$	1 C 3 H 1 C 1 O 1 O 1 H	2 (NH$_4$) 1 (SO$_4$)
1 O	Total Atoms	Total Atoms	2(NH$_4$)
	1 Ca	2 C	2 N
	2 C	4 H	8 H
	6 O	2 O	(SO$_4$)
			1 S
			4 O
			Total Atoms
			2 N
			8 H
			1 S
			4 O

$2 \times 3 = 6$ $\qquad\qquad$ $2 \times 4 = 8$

2 CO$_3$ $\qquad\qquad$ 2 NH$_4$

O━━━━ Molecular weight and formula weight are used interchangeably.

By knowing the mass of a mole and by understanding that a molecular formula mass can be determined by using the ratios of the atoms in the formula; we can begin to put together chemical equations. Chemical equations are the chemists' way of writing down the recipe to make a specific compound. For example, if we want to make sodium chloride, a chemist could react pure sodium with pure chlorine to get:

$$2\,Na + Cl_2 \rightarrow 2\,NaCl$$

In this form, the numbers in front of the formula indicates the number of moles required for the reaction. So, in this case we have 2 moles of sodium reacting with 1 mole of chlorine which will produce 2 moles of sodium chloride. By using the molecular weight, we can convert moles to mass and mass to moles. This allows chemists to measure the quantities of the reagents (the starting materials) and predict the overall yield based on the chemical equation. (See Stoichiometry. Stoichiometry refers to the coefficients in the chemical equation.)

But, this only works, if our equations are balanced. This means that we have to have the same number of atoms of each type on each side of the equation. (See Balancing Equations)

In this laboratory, you are going to use nuts, bolts and washers to represent atoms in a chemical equation. This will allow you to visually and physically investigate the concept of moles, molecular/atomic weight and predicting yields.

Stoichiometry

Recipe for Soda Bread

3 cups	Self Rising Flour (SFR)
3 tbsp	Sugar (S)
1 can	Club Soda (CS)

In a large mixing bowl, measure and combine the self rising flour and sugar. Mix using a spoon or other utensil. Slowly add the club soda to the flour mixture, and mix until all of the dry material has been incorporated into the batter. The batter will now have the consistency of cottage cheese.

Pour mixture into a greased bread pan. Bake in a preheated 375 degree F oven for 1 hour. Test with tooth pick or bamboo skewer to determine if the center done. (Test is conducted by sticking the skewer into the bread and when removing the skewer it comes out clean.

You can write this as a chemical equation:

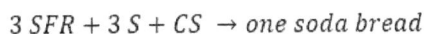

$$3\,SFR + 3\,S + CS \rightarrow one\ soda\ bread$$

Or, if we think that a cup represents a unit of self rising flour, a tablespoon is a unit of sugar, and a can is a unit of club soda; you could say that the recipe is 3 units flour, 3 units sugar, and 1 unit of club soda.

Chemical Equation for Combustion of Propane

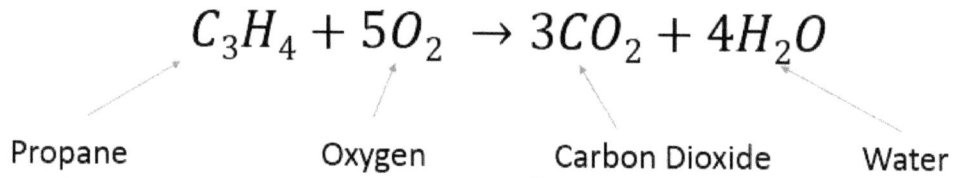

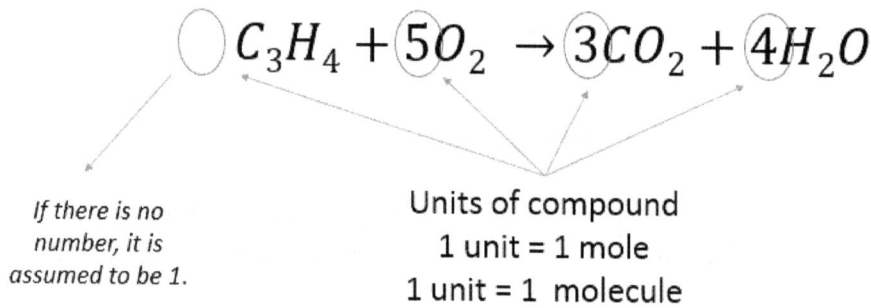

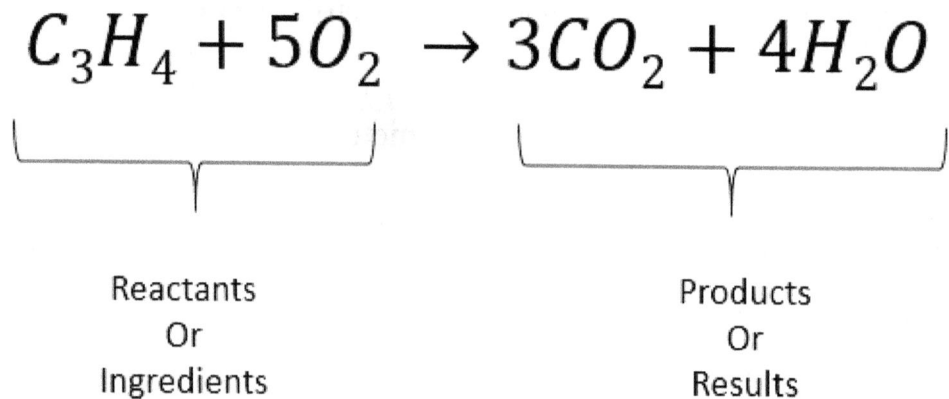

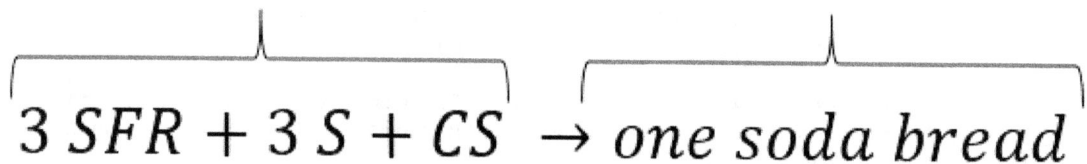

$$C_3H_4 + 5O_2 \rightarrow 3CO_2 + 4H_2O$$

Propane   Oxygen   Carbon Dioxide   Water

$$C_3H_4 + 5O_2 \rightarrow 3CO_2 + 4H_2O$$

If there is no number, it is assumed to be 1.

Units of compound
1 unit = 1 mole
1 unit = 1 molecule

$$C_3H_4 + 5O_2 \rightarrow 3CO_2 + 4H_2O$$

Reactants
Or
Ingredients

Products
Or
Results

$$3\,SFR + 3\,S + CS \rightarrow one\ soda\ bread$$

Balancing Equations

$$C_{12}H_{22}O_{11} + O_2 \rightarrow CO_2 + H_2O$$

Sucrose   Oxygen   Carbon Dioxide   Water

Reactants   Products

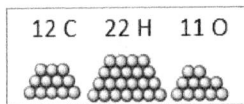

2 O

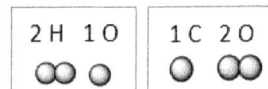

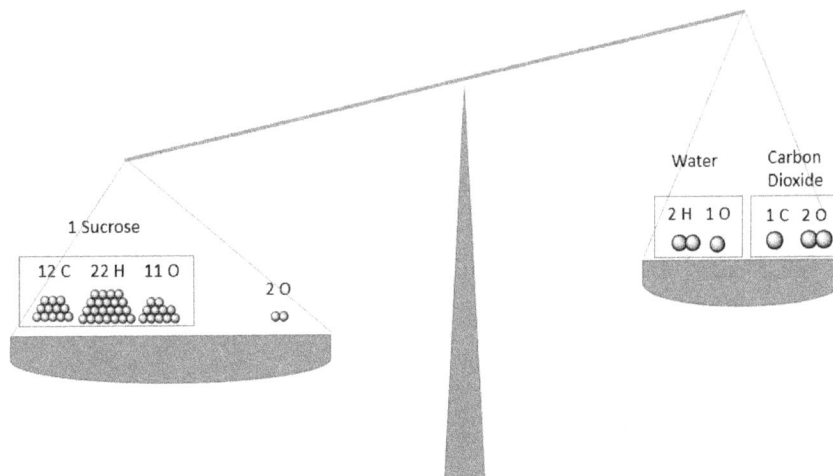

$$C_{12}H_{22}O_{11} \; + \; O_2 \; \rightarrow \; 12 \; CO_2 \; + \; H_2O$$

Sucrose   Oxygen   Carbon Dioxide   Water

Reactants   Products

12 C   22 H   11 O

2 O

2 H 1 O

1 C 2 O   1 C 2 O   1 C 2 O

1 C 2 O   1 C 2 O   1 C 2 O

1 C 2 O   1 C 2 O   1 C 2 O

1 C 2 O   1 C 2 O   1 C 2 O

Step 1 – Balance the Carbons

$$C_{12}H_{22}O_{11} \; + \; O_2 \; \rightarrow \; 12 \; CO_2 \; + \; 11 \; H_2O$$

Sucrose   Oxygen   Carbon Dioxide   Water

Reactants   Products

12 C   22 H   11 O

2 O

2 H 1 O   2 H 1 O   2 H 1 O

2 H 1 O   2 H 1 O   2 H 1 O

2 H 1 O   2 H 1 O   2 H 1 O

2 H 1 O   2 H 1 O

1 C 2 O   1 C 2 O   1 C 2 O

1 C 2 O   1 C 2 O   1 C 2 O

1 C 2 O   1 C 2 O   1 C 2 O

1 C 2 O   1 C 2 O   1 C 2 O

Step 2 – Balance the Hydrogens

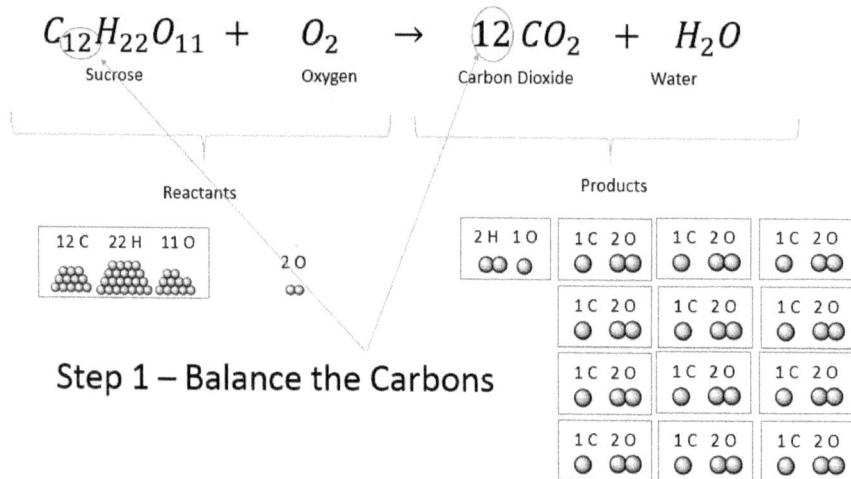

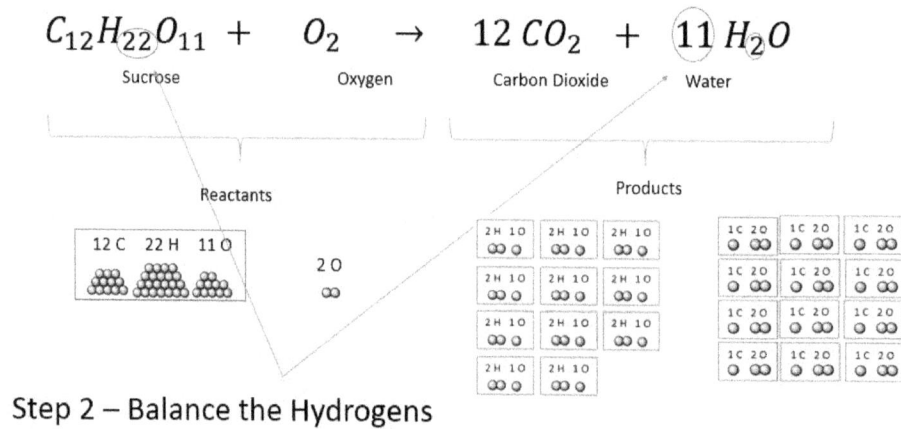

12 C   *Plus*   24 O from CO_2
22 H   11 O from H_2O
                          35 Oxygen total

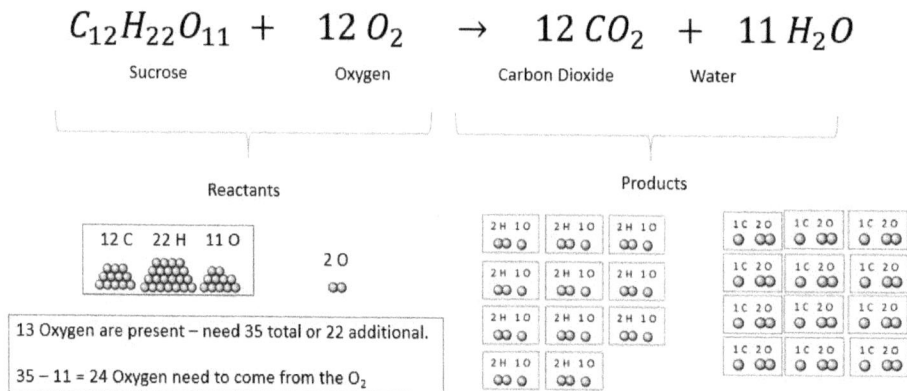

$$C_{12}H_{22}O_{11} \; + \; 12 \; O_2 \; \rightarrow \; 12 \; CO_2 \; + \; 11 \; H_2O$$

Sucrose   Oxygen   Carbon Dioxide   Water

Reactants   Products

12 C   22 H   11 O

2 O

13 Oxygen are present – need 35 total or 22 additional.

35 – 11 = 24 Oxygen need to come from the O_2

2 H 1 O   2 H 1 O   2 H 1 O

2 H 1 O   2 H 1 O   2 H 1 O

2 H 1 O   2 H 1 O   2 H 1 O

2 H 1 O   2 H 1 O

1 C 2 O   1 C 2 O   1 C 2 O

1 C 2 O   1 C 2 O   1 C 2 O

1 C 2 O   1 C 2 O   1 C 2 O

1 C 2 O   1 C 2 O   1 C 2 O

Step 3 – Balance the Oxygen

12 C   *Plus*   24 O from CO_2
22 H   11 O from H_2O
                          35 Oxygen total

$$C_{12}H_{22}O_{11} + 12\,O_2 \rightarrow 12\,CO_2 + 11\,H_2O$$

Sucrose   Oxygen   Carbon Dioxide   Water

Reactants   Products

12 C		12 C	
22 H	Plus	22 H	
11 O	24 O =	35 O	

12 C
22 H
35 O

Step 4 – Show Balance

General Safety Considerations for this Activity

During this experiment, we will be using a scale, nuts, bolts, and washers. There is no particular safety hazard associated with this activity. However, one may wish to observe cautions to prevent a choking hazard.

Equipment required for the Activity:

- Nut, bolt and washer set (You are going to need at least 8 nuts, 8 bolts, and 16 washers. It does not matter the size; however, the nuts, and washers should be able to be assembled on to the bolt. It is recommended that the size should be relatively small, such that the weight of 8 the same size bolt is approximately 40 to 50 grams, and the total weight is less than 100 grams.)
- Laboratory notebook
- Calculator
- Scale

Objectives:

- Visualizing a mole
- Exploring the stoichiometry
- Predicting the theoretical yield
- Understanding of limiting reagents

Procedure:

1. Read through all of the procedure and prepare your laboratory notebook.

2. Gather your materials, and develop your nut, bolt and washer table. For this experiment, we are going to use a made up unit of measure called the minimole. A minimole is much like a mole but instead of being 6.022×10^{23} of something, a minimole is 8 of something. So count out 8 of your nuts, 8 bolts, and 16 washers.

3. Measure the atomic weight in minimoles for each set and record the grams/minimole. Also, calculate your grams per item. (Remember a minimole is 8 of each item.)

	Bolts	Nuts	Washers
g/minimole			
g/item			

4. Prepare for your reaction. In this reaction, you will react one minimole of nuts, with one minimole of bolts and two minimoles of washers to get a minimole of widgets. (One widget is a nut, 2 washers, and a bolt. See picture) Assume that N is the symbol for nuts, B is the symbol for bolts and W is the symbol for washers. Then the molecular formula for a widget would be NBW_2.

5. Write out the equation for your reaction.

6. Using the data from your table, calculate the mass of one minimole of widgets and the mass of one widget.

7. Build your widgets.

8. Using the scale, measure the mass of one minimole of widgets and the mass of one widget.

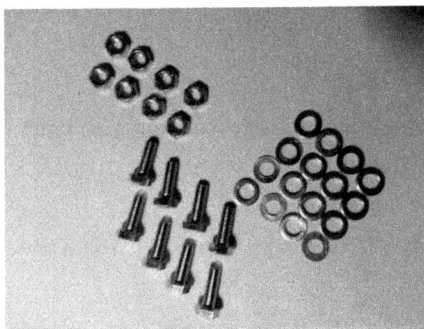

Nut and Bolt Set

Constructed Widget

Procedure (Continued):

9. Calculate the percent error between the predicted mass of the widget and one minimole of widgets, and answer the questions.

   a. Was there an error? Why or why not?
   b. If you had a mole (6.022×10^{23} of each bolts, nuts, and washers) of each, what would be the mass of the bolts, nuts and washers be? Compare that with the mass of the Earth 5.97×10^{24} kg.

Predicting Yield and Determining Limiting Reagent

10. Take apart your widgets.

11. Determine the mass of 10 washers.

12. Using the mass of the 10 washers, measure out a number of bolts, and nuts closest to the same mass. It will probably not be exact as you are not going to be able to have ½ a nut or bolt. Complete the following table in your notebook:

	Bolts	Nuts	Washers
Mass of 10 washers			
Number			
Mass (g)			

13. Calculate the number of minimoles that are present. (Take the grams and divide by grams/minimole.)

	Bolts	Nuts	Washers
Mass of 10 washers			
Number			
Mass (g)			
minimoles			

Procedure (Continued):

14. Answer the following: Does the number of minimoles make sense given that there are 8 items to a single mole? (Hint: What is the value for 10/8?)

15. Using your chemical equation, how many widgets would you predict that you can make?

16. Construct widgets until you run out of a specific ingredient. This ingredient is called the limiting reagent because when it is completely utilized the reaction stops. How many widgets were you able to construct?

Summing It Up:

Having worked through the laboratory exercise, hopefully, you can see how mini-moles can be a surrogate for moles. When you are working in the laboratory, you will be measuring out reagents, and determining the theoretical yield based upon the number moles that are present. You will determine limiting reagents, i.e. those reagents that will be used up first, thus stopping the reaction. These future laboratories will use calculations very similar to the ones used here.

Laboratory #8
Impacts on Physical Properties

Introduction

In the previous laboratory, you explored the two physical properties of materials, the melting/freezing point and the boiling point. As you observed, these properties describe when a material changes state, i.e. from a solid to a liquid or from a liquid to a gas. Additionally, while conducting the measurement of the physical property, you had a chance to observe that the addition of salt to the water bath made a difference in the physical properties of the water bath compared to what you observed with pure water.

Recall from the last laboratory, the water bath at the beginning of the experiment was much colder than what we know to be the freezing point of water. So, why did the water not freeze in the water bath? Additionally, you were able to get the water to boil in your pan at approximately 100 degrees C, but the water in the water bath mixed with salt did not boil at the same temperature. Why?

Based on what you know from a previous laboratory, the water in your pan (the pie plate) was not exactly the same as that in the water bath. You added salt. Based upon your observations, you can develop a hypothesis that can be tested:

>**Observation:** Salt impacts the boiling and melting point of water.

>**Hypothesis:** Washing soda may have a similar impact on the boiling and melting points of water.

During this course, although it has been inferred, we have not yet discussed the scientific method. You have been exposed to the difference between an activity and an experiment. You have begun utilizing some basic scientific skills: note taking, measurement, observation, and analysis. You have begun to learn the terminology of science. But, in this laboratory, you can begin to put some of those skills together and see how they relate.

In order to do this, you need to know the steps of the scientific method. The scientific method is a process. It is a formalized means of testing ideas and theories to explain what you have observed in the laboratory. In many texts, the method is listed as a set of steps:

1) Develop a hypothesis

2) Test the hypothesis

3) Analyze the data to determine whether the hypothesis is correct or needs to be modified.

4) Develop new/modified hypothesis

5) Repeat

But this is a simplified version of what really happens. Think about it?

Do you really start out with a hypothesis or question?

When testing the hypothesis, have you accounted for all possibilities?

Where does preparation and the scientific body of work fit into the process?

The scientific method really starts with an observation, much like the one that you had in the previous laboratory: salt impacted the boiling and melting points of water. Now you have a question – why or how did salt impact the boiling and melting points?

Once you have an initial question, you may find yourself asking more:

If salt can impact the boiling and melting points,
what about other materials?

Is water the only material this happens with?

Let's think of this process in terms of the laboratory you are about to conduct, there will be three questions you will be investigating:

1) How does salt impact the boiling and melting points?

2) Is there a relationship between the amount of salt added and the observed impact?

3) Do other substances like washing soda have a similar impact?

These are questions you want to investigate – but they are not hypotheses. For a hypothesis to be "scientific" it must be testable, and in principle be capable of being proven wrong. In order to develop a good hypothesis to test, you may have to do a bit of background research. This is called a literature search, i.e. looking up information from previous scientific experiments. This will be one step in your laboratory procedure. But, to help us in the discussion here, let's look at the following question using an Internet browser:

"How does salt impact the boiling point of water?"

In the Internet search, one of the first items that appears is a brief article on "eHow". Its first paragraph states:

> Adding salt to water raises the boiling point. This is a scientifically measurable effect, but it takes 2 oz. of salt to raise the boiling point of 1 liter nearly 2 degrees F, so the effect is not noticeable in your average kitchen. Salt added to water also lowers the freezing point. (Moore, 2014)

Using this information, you can refine your question into a true hypothesis:

Adding salt to water will raise its boiling point and lower its melting point.

Hopefully, this corresponds to the observations that you made in the previous laboratory. Let's look at the hypothesis statement, is it testable? Yes, we can add salt and see if the boiling point raises and the melting point lowers.

Through your research you will probably find a number of different experimental methods or procedures you can use to test your hypothesis. This laboratory will provide you with one.

Now, you are ready to move through the scientific method by conducting the experiment. And, see if the data that you collect verifies or if you have to make a modification to your hypothesis and/or your experimental procedure.

There is another step that is left out of our simplified scientific method, communicating your results. If scientists did not communicate their results, how would you refine your question into to a hypothesis and/or develop your methodology to test it? Thus, this is a critical step that we will explore later in this course.

For this particular laboratory:

In this laboratory you are going to measure the impact of salt and sugar on the freezing of water, or more formally you are going to investigate the freeze point depression of water. And, you know that you will be doing a bit of background research as to why this happens. You are likely to find out the following:

Salt and washing soda will be considered solutes
(i.e. they are dissolved into the water, thus water is the solvent)

Through experimentation, the observed impact on the boiling point or freezing point can be written as a mathematical expression:

$$the\ change\ in\ temperature\ (\Delta T) = Km$$

Where K is a constant related to the properties of the material, Kb represents the boiling point elevation constant, and Kf is the freezing point depression constant. These constants can be looked up in reference tables. The m is the molality or the number of moles of solute per kilogram of solvent.

Kb of water is 0.512 °C/m and Kf of water is 1.86 °C/m

General Safety Considerations for this Activity

For this experiment you will be using:

- Water
- Ice
- Salt
- Washing Soda

If you haven't already, obtain a SDS for each of these materials and review the potential hazards associated with them.

As will all of the laboratories, you should wear your goggles and use gloves while conducting the experiment.

Be sure to complete a hazard analysis before you begin.

Equipment required for the Activity:

- Water
- Ice
- Salt (approximately 15 grams of non-iodized salt)
- Washing Soda (approximately 10 grams)
- A two-cup measuring cup with ml markings
- 2 Styrofoam cups
- Scale
- Thermometer
- Laboratory Notebook

Objectives:

- Determine freezing point depression of water
- Determine K_f for water

Procedure:

Preparing for the experiment:

1. Begin to set up your laboratory notebook for this activity. Record the title, date, etc. Your first entry should be the background information that you obtain from the introduction to this laboratory and your "research" that you will be conducting in Step 2.

2. As mentioned in the introduction to this laboratory, take a few minutes to do some background research about freeze point depression. See what information that you can find.

3. Read experimental procedure. While reading over the procedure, you need to be looking for items that may need to be included in your hazard analysis, and you need to be observing what type of data that you will be collecting.

4. Complete your hazard analysis.

5. Develop your hypothesis for this laboratory. Write this hypothesis in your laboratory notebook.

Procedure (Continued):

6. In your laboratory notebook, outline the experimental procedure that you will be using.

7. Begin your data collection section, i.e. create a data collection table. You can use the following table or you can create your own:

Raw Data for Salt				Raw Data for Washing Soda			
mass of empty cup				mass of empty cup			
mass of cup plus water/ice				mass of cup plus water/ice			
mass of water				mass of water			
	Temp. (°C)	Grams of Salt Added	Total Grams of Salt in Solution		Temp. (°C)	Grams of Washing Soda Added	Total Grams of Washing Soda in Solution
Initial (T$_i$)				███			
First Addition (T$_1$)				███			
Second Addition (T$_2$)				███			
Third Addition (T$_3$)				███			
Observations							

8. Gather your equipment and prepare to conduct the experiment.

Procedure (Continued):

Experimental procedure:

To prepare for this experiment – fill your two-cup measuring cup with ice (crushed is preferred), and add water to the two-cup line. This is your "stock" water for the experiment. Note: you can use tap water for this experiment.

1. Weigh one of the Styrofoam cups. Record the mass of the cup.

2. Add to the Styrofoam cup, approximately 50 to 75 grams of water mixed with ice from your stock water/ice solution. Weigh the cup with the water/ice mixture.

3. Allow the cup and water/ice mixture to equilibrate (i.e. the temperature no longer changes) and measure the temperature.

4. Measure out approximately 5 grams of salt. (Record the actual mass.) Add the salt to the water/ice mixture and stir with a stirring rod or spoon. Try to get as much salt to dissolve as you can.

5. Record the temperature of the salt/water/ice mixture when the temperature appears to stabilize, i.e. there is no real change in temperature.

6. Measure out approximately 5 more grams of salt. (Record the actual mass.) Add the salt to the salt/water/ice mixture and stir with a stirring rod or spoon. Again try to get as much salt to dissolve as you can.

7. Record the temperature of the salt/water/ice mixture when the temperature appears to stabilize, i.e. there is no real change in temperature.

8. Repeat Steps 6 and 7.

9. Record any additional observations. For example: were you able to get the salt to completely dissolve?

10. Set salt/water/ice mixture aside.

11. Starting with the second dry Styrofoam cup, record the mass of the cup.

12. Add to the Styrofoam cup, approximately 50 to 75 grams of water mixed with ice. Weigh the cup with the water/ice mixture.

13. Repeat Steps 3 through 9, except instead of adding 5 grams of salt, add 3 grams of washing soda. And, document the temperature readings.

Procedure (Continued):

14. Once you have collected the data for both the salt/water/ice and washing soda/ water/ice solutions, pour the mixtures down the sink. And, clean up the area.

Doing the Calculations:

1. Do the initial calculations to complete your data table. Find the mass of the water, and the total grams of solute (salt or washing soda added).

2. Prepare a table in your laboratory notebook to summarize your calculations. See Example Table.

2. Calculate the change in temperature between the initial reading and the reading after the addition.

3. Calculate the molality of each solution. Recall from the introduction that molality is the number of moles of the solute divided by the kilograms of solution. Also recall to calculate moles, you need to take the number of grams of the compound divided by the molecular weight. So your first step here is to calculate the molecular weight of salt and washing soda, then calculate the moles.

	Calculations for Salt				Calculations for Washing Soda			
	MW of Salt				MW of Washing Soda			
	Mass of water in Kg				Mass of water in Kg			
	Change of Temp from initial reading w/no Solute	Moles of Solute	m	K_f Calculated (°C)	Change of Temp from initial reading w/no Solute	Moles of Solute	m	K_f Calculated (°C)
Initial								
First Addition (T_1-T_i)								
Second Addition (T_2-T_i)								
Third Addition (T_3-T_i)								

Procedure (Continued):

4. Find the average for the freezing point depression constant.

5. Compare your findings with each other and the literature value of 1.86 °C/m.

Summing It Up:

You will want to answer the following questions in your laboratory notebook:

- Did your results support or contradict your hypothesis?

- How did your values for K_f compare with the literature value?

- If your value disagreed with the literature value, why do you suppose this to be the case? (Hint: what are your sources of error?)

- How might you adapt this experiment to conduct a study for boiling point?

Laboratory #9
Writing a Laboratory Report

__Introduction__

In two of the previous laboratories, you observed that the addition of salt or another ionic compound could impact the boiling and melting points of water. In the last laboratory, you learned about developing and testing a hypothesis. Now, it is time to add another tool to your scientific tool kit, the laboratory report.

A laboratory report, while used to present your results from as structured laboratory exercise, is a preliminary form of a technical scientific paper. Scientific papers are used by chemists and other scientists to communicate their findings in the laboratory to support or disprove various hypotheses and/or scientific theories. The purpose of the paper is to outline the development of the hypothesis to be tested, provide other scientists information about the experiment performed, communicate the data collected, and provide some analysis as to how this data supports or disproves the presented hypothesis. This information is then used by other scientists to either reproduce the experiment performed or to use the results to develop their own questions or hypotheses.

So in this activity, you are going to use the information that was developed to generate, a laboratory report. Laboratory reports are used by instructors to help prepare you to write your own scientific paper at some point in your career, but it also helps to enhance other critical skills. These skills include: accurately communicating information about the experiment or procedure you performed, providing information about the subject you are studying, and data analysis, i.e. what is it you learned from this experiment or activity. Laboratory reports are used in many science laboratory classes; biology, chemistry and physics. So, developing good writing skills will help you in many of your future courses and will help you in your future job as well.

The laboratory report is primarily an archival report. A scientific paper may be archival or persuasive. In a persuasive paper, the author is trying to persuade others that the experiment performed and the data obtained support a particular position. You may use this persuasive type of approach for a science fair project to show that your data supports a particular hypothesis.

Laboratory Report

Title
Should describe the laboratory in approx. 10 words.

Abstract
The purpose is to allow the reader or user to determine if it suits their needs. It is a summary of the overall laboratory. Even though it comes at the top – it should be written last. It should be concise, 200 words or less.

Introduction
This section of the report should answer the questions:
Why was the laboratory performed?
What is known about the subject?
What was specifically investigated?

Equipment and Procedure
This section of the report should list the equipment needed and the laboratory set-up. Then detail how the experiment or laboratory was performed so that it can be repeated by someone else.

Results
This is a presentation of the data gathered as a result of the experiment.

Discussion and Conclusions
In this section the following questions should be answered:
Did the experiment/laboratory performed as planned?
What conclusions did the data provide?
If the data did not support the hypothesis presented in the introduction, why or why not?
What changes or other controls might be included in the next experiment?

References and Other Information
This section should include any references that were used in the laboratory report. It may also include information about how to obtain laboratory equipment, acknowledgements of individuals who provided assistance.

This is different than an archival approach, which is generally written to document the experiment performed, the data obtained, and any factors that may have been present to put the data in question. These types of reports/papers are used to document historical data such as the gallons of milk provided by a herd of dairy cows per year under specific weather and feeding conditions. This information is very valuable to others when setting up experiments in the future. These types of approaches are used by chemists in the areas of chemical synthesis (how to make a specific molecule) as well as process chemistry. Your laboratory report is an archival report as it is documenting how you performed the laboratory and communicates your results. You are not trying to persuade anyone about how your results change a theory.

In either case, the laboratory report or scientific paper follows a specific format. It includes:

- A Title
- An Abstract
- An Introduction
- A section on equipment and procedures
- Results
- Discussion
- References and other information

The title should readily describe the experiment or laboratory performed. It should be short (although as science has grown the title length has grown as well). For most laboratory reports, try to keep your title to less than 10 words. A good title will make it easy for the reader to understand. For example, if you were writing a laboratory report on the density experiment you did earlier in the semester, your title might read:

Calculating the Density of Various Objects

The abstract is a brief concise summary of your experiment and results. In a scientific paper, the abstract is used by other scientists to allow them to see if the information contained in the paper is relevant to their needs. For example, a scientist may be working on a project to determine whether or not increasing the temperature of the starting resin mixture makes a stronger fiber. This scientist will be looking for papers by others that explores factors that showed an increase in fiber strength. By reading the abstract, the investigator can go through a large number of papers on making of fibers to determine which ones should be read more closely. While the abstract is located at the beginning of the laboratory report or scientific paper, it should be written last as it is a summary of the paper. Abstracts should be about a paragraph and less than 200 words.

The introduction is the portion of the report that provides the background to the overall report. It should contain information about why the laboratory or experiment is being performed. This section of the report also details the background information that is used to develop the hypothesis being tested as well as what is specifically being investigated. Let's go back to the density laboratory. If you were writing a laboratory report on that activity, your introduction should include some information about what is density and how it is determined. Then describe why you can't always determine density by a simple measurement of the mass and calculating the volume.

The next section, equipment and procedures or it may be referred to as materials and methods, details what is needed to perform the experiment or laboratory and how it is to be completed. Each of the laboratories in this manual have an equipment and procedure section. When you are writing your laboratory reports, you need to write your equipment and procedure section in your own words.

Safety

So far, there has been no mention of where to write about safety considerations. And, if you look on the Internet about how to write a laboratory report, safety is typically left out as well. Safety should be discussed in the equipment and procedures sections. Be sure to include information about the hazards of the materials to be used and note any potential hazardous steps during your discussion of the procedure.

The results section contains information about the data you obtained. It usually includes data tables, graphs, photographs, and figures. The information should be clear and concise. Figures, tables, graphs, and photographs should be titled, labeled and captioned to make it clear to the reader what information is being presented. In providing the text for this section, you focus on general trends and observations.

The analysis of your data is provided in the discussion section of the report. The discussion section of the report should not be a restatement of the results section, but should focus on why the trends either support or disprove the hypothesis or objective outlined in the introduction. It is in this section where the types of errors or problems encountered should be discussed. Suggested modifications or future experiments are also included here.

Finally, recognition to the work of others has to be included in your report. This is done in the reference section. Here you will document the sources of materials used in writing your laboratory report. Additionally, you may want to include supplementary information about where one might obtain specific pieces of equipment or the source of your laboratory samples. Using the appropriate bibliographic citations is a must.

With this brief explanation and format, you should now be able to try your hand at your own laboratory report.

General Safety Considerations for this Activity

This activity is the writing of a laboratory report on the "Impacting Physical Properties." There is no particular safety hazard associated with this activity. However, be sure to include safety information in the laboratory write-up.

Equipment required for the Activity:

- Laboratory notebook with results from previous laboratory
- Writing materials
- Internet access

Objectives:

- Writing a laboratory report

Procedure:

1. Go online and obtain two scientific papers. (You can use an Internet search or Google Scholar) It doesn't matter if you understand all of the material in the papers, what you will be looking for is how the papers are organized. In your laboratory notebook, document the following for each paper:

The Title

Did the paper have an abstract?

Did the paper have an introduction?
　　　(Was it titled something else? If so, what did the author call it?)

Was there an equipment and procedure section? Was it one or two sections? Did it have a different title?

Did the paper include a results section? Was it by itself or combined with the discussion section?

Were references included?

Was other information included?

Procedure (Continued):

2. Using the information provided in the introduction, answer the following about your scientific papers:

> Were the papers similar in outline to that described?

> What were the major differences you observed?

3. Using the information provided in the introduction to this activity and from the "Impacting Physical Properties" laboratory, write a laboratory report.

Summing It Up:

After you have completed your laboratory report, make a journal entry into your laboratory notebook. In this journal entry discuss the following:

> 1) What did you like about this activity?

> 2) Having completed the laboratory report, what things might you do different in your next laboratory report?

Laboratory #10
Exploring Solubility

Introduction

As you have been exploring physical properties, you have probably noticed that the temperature of the solvent, in most cases water, impacted how much material was able to be dissolved in a given amount of water. What you were observing was how temperature impacted the solubility, the quantity of a substance that will dissolve in a given amount of solvent, of the material.

If you have ever made sweet tea, Kool Aid™, or tried to sweeten iced tea, coffee, or hot tea; you have experienced the effects of temperature on the solubility of sugar. The solubility of a solid can be impacted by a number of different factors, but for ionic solids (salts) the solubility can be directly impacted by temperature. Because, solubility of an ionic solid is impacted by temperature, solutions are described as saturated, unsaturated, or supersaturated. In a saturated solution, no additional solute (salt) can be dissolved into the mixture at that temperature. In an unsaturated solution, additional material can be dissolved. In a supersaturated solution, more solute is dissolved than is expected. Southern sweet tea is an example of a supersaturated solution.

Solubility is expressed in grams of solute (the solid salt) per quantity of solvent (typically in 100 grams) at a specific temperature. The chart below shows the solubility of some common salts as a function of temperature. As you can see from the chart, table salt has a fairly flat curve and as you increase in temperature there is little difference in the amount of salt that can be dissolved. However, there are significant differences in the amount of material that can be dissolved when using potassium nitrate (a type of salt that is found in fertilizers).

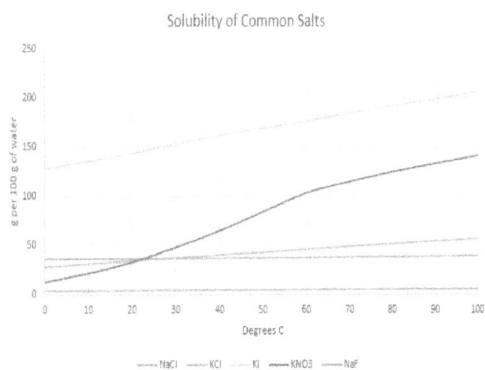

Solubility of Common Salts

In this laboratory, you are going to study the solubility of calcium chloride in water and develop your own solubility curve. By using the saturation point, i.e. the point when the material crystallizes and falls out of solution, you will be able to determine the solubility at a given temperature.

General Safety Considerations for this Activity

This laboratory requires the use of calcium chloride, water, and a heat source. As calcium chloride is used in the making of cheese, wine, and beer; it is not a particularly hazardous substance. However, as you have learned during your reviews of the SDSs, most materials have some level of hazard. The SDS for calcium chloride lists the following hazards: an inhalation dust hazard associated with large quantities, ingestion of high amounts may result in gastrointestinal problems (nausea and vomiting), eye or skin irritation associated with the chloride or from mechanical abrasions. Thus, safety goggles and gloves are recommended when conducting this laboratory.

Additionally, since heat is required and you will be working in a range of temperatures between 30 and 80 degrees C, there is a possibility of burns from the hot water. Care should be taken while working at these temperatures. Hot pads or mitts may be useful and test tube holders will be helpful.

Equipment required for the Activity:

- Calcium chloride
- Water (distilled)
- Heat Source (stove)
- Water bath (Sauce pan filled with water or a beaker with water that can be placed over a Bunsen burner or can be placed on a laboratory heating element.)
- Test tubes (4)
- Test tube holder and tongs
- Scale
- Stirring rod
- Thermometer
- Plastic Spoon
- Laboratory Notebook
- Weighing tray (wax paper or foil)
- Marker
- Graduated cylinder

Objectives:

- Exploration of solubility versus temperature
- Development of a solubility curve for calcium chloride

Procedure:

1. Obtain a copy of the safety data sheet for calcium chloride. Record information related to calcium chloride in your laboratory notebook. Recall, the information you have been recording includes:

- The common name
- The trade name or product specific name
- Chemical name
- Chemical formula
- What the material is used for or common use
- Warnings/Cautions associated with the material
- Any special handling instructions

2. Read through the entire laboratory before starting. Complete your hazard analysis, and document this in your laboratory notebook.

3. Set up your laboratory apparatus.

If you were in a true laboratory, your apparatus would look like:

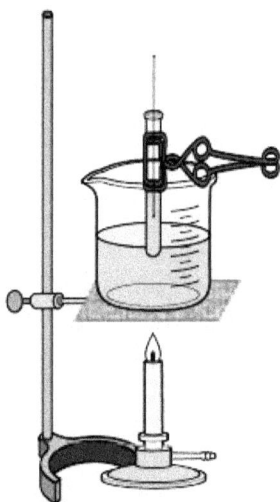

However, as you are most likely going to be conducting this experiment at the kitchen stove, you will need to fill a small to medium sauce pan approximately two-thirds full. The water should be high enough to allow you to heat up the test tubes which will contain water and calcium chloride. This water bath will be similar to the one you used in your melting point laboratory.

Obtain 4 test tubes. Label the test tubes 1, 2, 3, and 4.

Procedure (Continued):

4. Measure out a quantity of the calcium chloride between 8 and 9 grams, and place it in the test tube marked 1. Repeat, using a quantity approximately 1 gram more than the previous measurement and place in the test tube marked 2. Repeat, for test tubes 3 and 4. You should now have four test tubes with a differing amount of calcium chloride. Be sure to record your actual amounts in a data table in your laboratory notebook.

Test Tube	Mass of CaCl2 (g)
1	
2	
3	
4	

4. Add 10.0 mL of distilled water to each test tube. (Note your thermometer must be able to touch the water level in the test tube.)

5. Begin heating the water in the water bath. Heat the water until the water bath reaches a temperature of 90 degrees C. Adjust the heat to maintain your water at about this temperature.

6. Using your test tube holder, place the test tube with the largest amount of calcium chloride of the test tubes into the water bath. Stir the calcium chloride mixture with a stirring rod until all of the calcium chloride has been dissolved. Note: depending on your source of calcium chloride, this step may take a bit of time, but eventually your calcium chloride will go into solution, i.e. has dissolved.

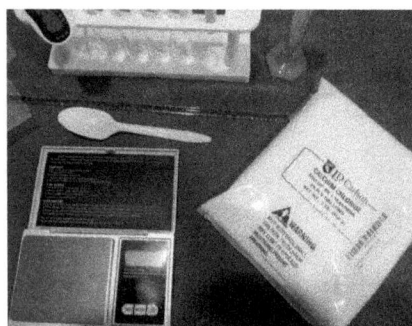

Experimental Equipment

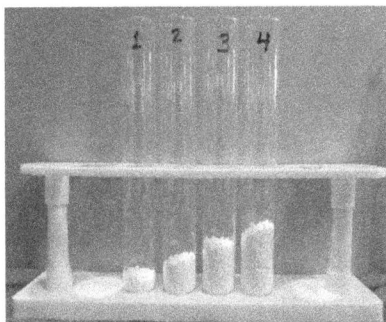

A

B

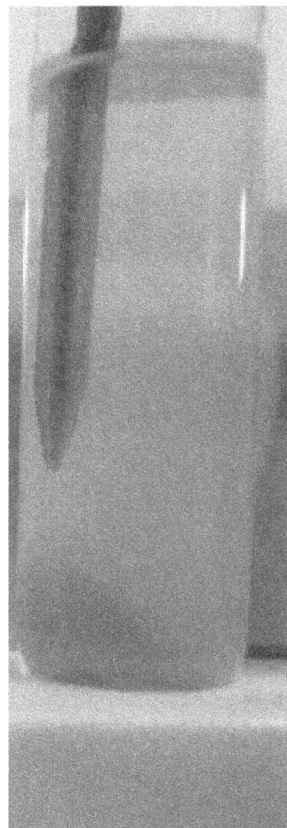

C

Images from different stages of the experiment.

A. Test tubes with differing amounts of the calcium chloride.

B. Test tubes with distilled water added prior to heating.

C. Test tube 4 after heating with all the material dissolved into solution. Metal tip is the thermometer.

Procedure (Continued):

7. Place the heated test tube in the test tube rack, remove the stirring rod, and replace with the thermometer.

8. Reduce the heat on the water bath.

Procedure (Continued):

9. Watch for the first signs of crystallization. When you see crystallization start, record the temperature in your laboratory notebook.

Test Tube	Mass of CaCl2 (g)	Temp. of Crystallization
1		
2		
3		
4		

10. Rinse the stirring rod and thermometer with distilled water. Repeat steps 6 through 9 for the remaining 3 test tubes.

11. Clean up your laboratory work space. Dispose of the calcium chloride/water mixture in the sink and rinse.

Summing It Up:

Using the proportions, grams of solute (calcium chloride) to quantity of solvent (mL of water), calculate the equivalent grams to 100 mL of solvent. Or, grams/5.0 mL is equivalent to x grams/100 mL. Also, calculate the grams/100 grams of solvent ratios, using the conversion factor for water of 1.00 g/mL. Record this in your notebook.

Plot your experimental results on a graph. The y-axis should be in grams of $CaCl_2$/100 grams of water. The x-axis will be in degrees C. Draw your solubility curve.

Answer the following questions in your laboratory notebook:

1. Using the Internet, find a solubility curve for calcium chloride. How does your solubility curve, compare with the curve you found?

2. Using your curve, predict what the solubility of calcium chloride will be at 30 degrees C?

3. How did the saturation point help you determine the solubility of the mixture?

4. What factors impacted your laboratory experiment?

5. What things would you change about this experiment?

Laboratory #11
Precipitation and Yield

Introduction

In the previous laboratory, you explored how temperature impacts the solubility of an ionic solid. As you might expect, there are other factors that can impact solubility. For example: surface area and pressure can have an impact on the amount of material that can be dissolved into the solvent. The characteristics of the solvent can also impact whether or not a material can be dissolved. For example, oil does not dissolve into water due to the nature of the two liquids. And, not all ionic solids will dissolve in water.

Recall in an earlier discussion, the difference between covalent/molecular versus ionic bonds were discussed. In a covalent/molecular bond, the electrons are shared between the atoms, while in an ionic bond, the electron is actually transferred between the atoms. So in the case of sodium chloride, the sodium atom gives up an electron to the chlorine atom. Thus, the sodium atom now has a positive charge while the chlorine atom is negatively charged. The interaction between the positively and negatively charged atoms is what holds them together, much like the attraction between two magnets. For some ionic compounds, these bonds are easily broken, allowing the solids to be readily dissolved in water. But, for others, the ionic bond is so strong it will not dissolve in water.

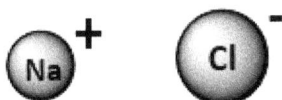

It turns out that solubility of ionic compounds has to be explored empirically. This means that the rules for solubility have been developed over a number of experiments. Because, solubility in water is more empirical, students of chemistry must learn the solubility rules. These rules are provided in the table. However, these rules are very helpful in predicting whether or not a precipitation reaction will occur, allowing for separation or removal of specific compounds, and for identifications of unknowns. (You will get an opportunity, to use these rules to help you identify an unknown in a later experiment.)

Solubility Rules

1) All common compounds of Group 1 A metal ions (Li^+, Na^+, K^+, Rb^+, Cs^+) and the ammonium ion $NH4+$ are soluble in water.

2) The common nitrates ($NO3^-$); acetates (CH_3COO^-); chlorates ($ClO3^-$), and perchlorates ($ClO4^-$) are soluble in water.

3) The common Cl^- are soluble in water EXCEPT $AgCl$, Hg_2Cl_2, and $PbCl_2$

4) The common Br^- and I^- have similar behavior to the chlorides, but there are some exceptions. As the halide ions increase in size, the solubilities slightly decrease.

5) The common $F-$ are soluble in water EXCEPT MgF_2, CaF_2, SrF_2, BaF_2, and PbF_2

6) The common sulfates, SO_4^{2-}, are soluble EXCEPT $PbSO_4$, $BaSO_4$, and $HgSO_4$, $CaSO_4$, $SrSO_4$, and Ag_2SO_4 are moderately soluble.

7) The common metal hydroxides, OH^-, are insoluble in water except with those of the Group 1A metals and the heavier members of the Group 2A metals, beginning with $Ca(OH)_2$.

8) The common carbonates, CO_3^{2-}, phosphates, PO_4^{3-}, and arsenates, AsO_4^{3-}, are insoluble in water except those of the Group 1A metals and NH^{4+}.

9) The common sulfides, S^{2-}, are insoluble in water except those of the Group 1A, Group 2A, and NH_4^+.

You should commit these rules to memory. However, remembering Rules 1, 8 and 9 — will get you through most problems.

In this laboratory, you are going to prepare two ionic solutions by dissolving two compounds into water. The compounds you will be working with are Epsom salt and washing soda. Epsom salt is magnesium sulfate, and washing soda is sodium carbonate. Looking at the solubility rules, you will notice that all sodium compounds are soluble in water, so the sodium carbonate should readily dissolve. Similarly, from the solubility rules, most sulfates are soluble in water. Thus, the Epsom salt should also dissolve into water. But, what is going to happen when you mix the two solutions together?

In order to predict what is likely to happen, let's look at a similar experiment. On the Internet, if you search for a precipitation reaction, you are likely to find a reaction between lead nitrate and potassium iodide. These two solutions are used because they provide a very visible reaction and the precipitate is yellow in color. The reaction that is occurring is called an exchange reaction because the cation (the positively charged ion) as well as the anion (the negatively charged ion) are being exchanged between the two compounds. And, when the exchange occurs, one of the compounds formed lead iodide is insoluble in water, and a solid forms that can be filtered out of the solution. Similarly, you are going to see the same thing happen with the solution of Epsom salt when mixed with a solution of sodium carbonate.

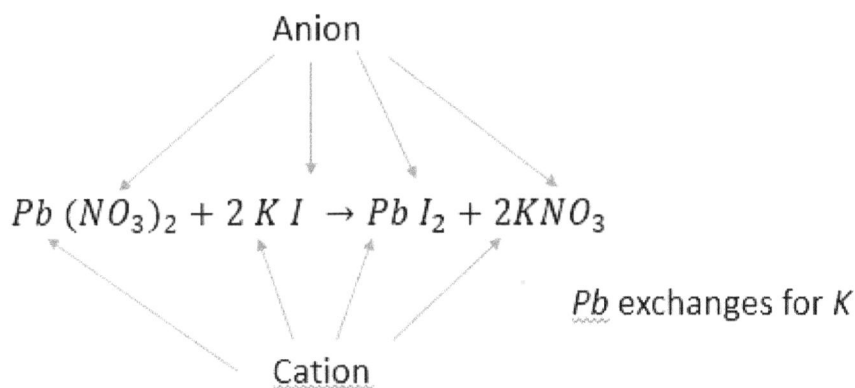

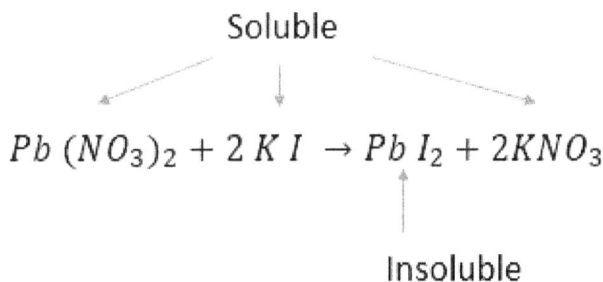

Anion

$$Pb\,(NO_3)_2 + 2\,K\,I \;\rightarrow\; Pb\,I_2 + 2KNO_3$$

Pb exchanges for K

Cation

Soluble

$$Pb\,(NO_3)_2 + 2\,K\,I \;\rightarrow\; Pb\,I_2 + 2KNO_3$$

Insoluble

Demonstration on a larger scale of the experiment to be performed in this laboratory.

General Safety Considerations for this Activity

In this laboratory you will be using washing soda and Epsom salt. Both washing soda and Epsom salt have ingestion and irritation hazards associated with them. Therefore, for this laboratory, you should be using gloves and goggles.

Equipment required for the Activity:

- Epsom Salt
- Washing Soda
- Distilled water
- Test tubes (3)
- Graduated cylinder
- Stirring Rod
- Coffee filter
- Scale
- Weigh paper or foil
- Laboratory notebook
- Scissors

Objectives:

- Preparing ionic solutions
- Filtering a precipitate
- Calculating theoretical yield
- Measuring actual yield
- Determining sources of error
- Identifying the limiting reagent

Procedure:

1. Using the Internet, locate the SDSs for Epsom salt and washing soda. Record information related to these compounds in your laboratory notebook. Recall, the information you have been collecting includes:

- The common name
- Chemical name
- Common use
- Any special handling instructions
- The trade name or product specific name
- Chemical formula
- Warnings/Cautions

Procedure (Continued):

2. Read through the entire laboratory before starting. Complete your laboratory hazard analysis, and document this in your laboratory notebook.

3. Complete the following prior to starting the laboratory:

Write out the balanced chemical equation for the reaction.

Calculate the molecular weight of both of the reagents and both of the products.

Using the solubility rules, which of the products is likely to be the precipitate?

4. Prepare the two ionic solutions. Using two test tubes, place 10 mL of distilled water in each test tube. Measure out approximately 1 gram of Epsom salt. Record the mass. Add the Epsom salt to one of the test tubes with water. Label this test tube so that you know which test tube has the Epsom salt. Measure out approximately 1 gram of washing soda. Record the mass. Add the washing soda to the other test tube with water. Label this test tube. (Note: You are going to have to allow the solids to completely dissolve, you may have to use a stirring rod to help this process.)

5. Prior to mixing the solutions, determine the number of moles of Epsom salt and washing soda you have available for the reaction. Answer the following:

Which reagent has the fewest moles?
Which reagent will determine the number of moles of product produced?
What is the predicted number of moles of the precipitate?
What is the predicted mass of the precipitate?

6. Pour the Epsom salt solution into the washing soda solution. Observe and document what happens.

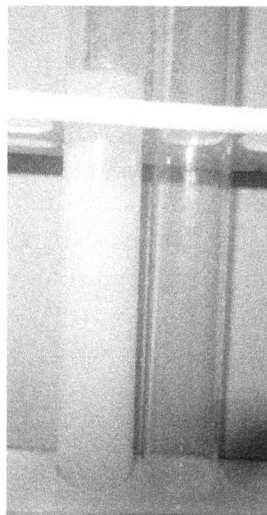

Test tubes after reaction.

Procedure (Continued):

7. Prepare a test tube for filtering. Cutting a coffee filter down to a usable size, make a cone filter for a test tube, similar to what you see pictured.

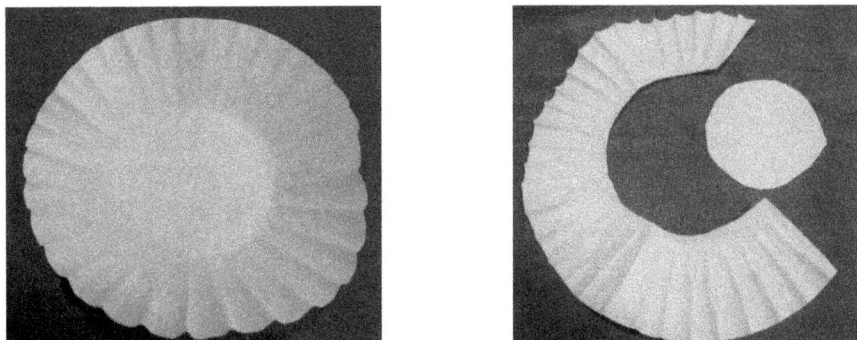

Cutting the coffee filter to obtain a circular filter.

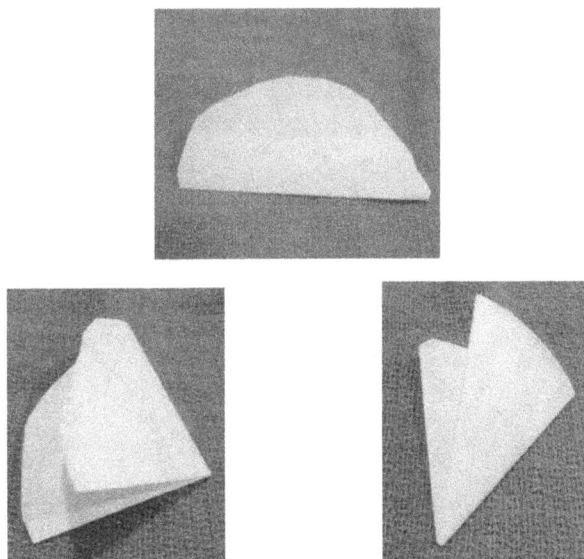

Folding the filter into a cone.

Procedure (Continued):

8. Record the mass of the filter.

9. Slowly pour the solution with the precipitate into the filter and allow to drain. If some of the precipitate remains in the test tube, use a bit more distilled water to wash it out and pour the material into the filter. (Note: This may take some time, be patient.)

10. Allow the filter paper with the precipitate to dry. (Note: it may take a couple of days for the filtrate to completely dry. Be patient.) Clean up your work space.

11. Once the filter paper is dried, record the mass of the paper with the precipitate.

12. Determine the mass of the precipitate.

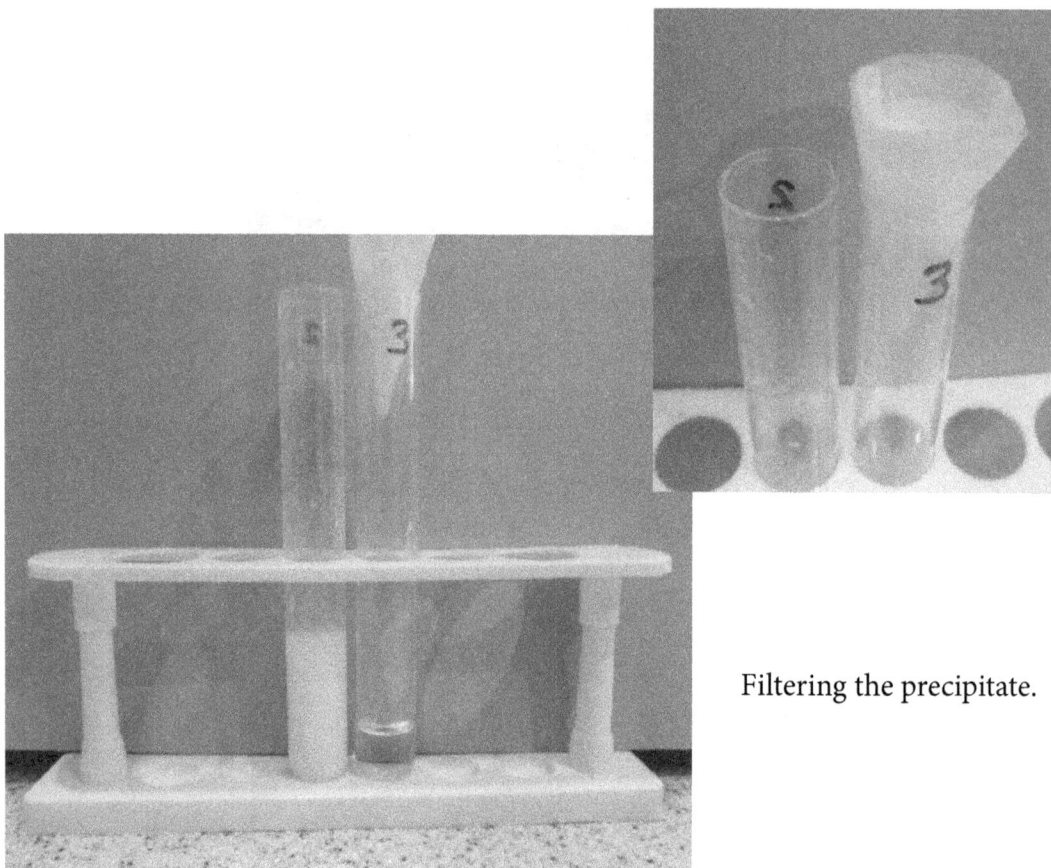

Filtering the precipitate.

Summing It Up:

Compare your predicted mass of the precipitate with that of your actual mass. (Use the percent error equation.) Document this in your laboratory notebook.

How did your experiment compare with your predictions?

What sources of error were present?

What would you have done differently?

Laboratory #12
Exploring Chemical Reactions

Introduction

In the previous laboratory, you had the opportunity to work with one type of chemical reaction, the double replacement reaction. Recall, in this reaction the cations and anions exchanged to form two new compounds, one that was soluble and one that was insoluble. A double replacement reaction has a general form:

$$AX + BY \rightarrow BX + AY$$

In this reaction, the cations and anions essentially change partners. But, this reaction only occurs, if one of the resulting compounds forms a precipitate. Otherwise, both the starting reagents and the products are present in the solution, i.e. the ions stay mixed.

So, by using your solubility rules, you can look at a number of reactions to determine whether or not a reaction of this type will occur. For example: What would have happened if we had used a table salt solution (sodium chloride) rather than the Epsom salt in the previous experiment? The chemical equation would have looked like:

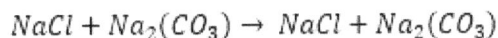

$$NaCl + Na_2(CO_3) \rightarrow NaCl + Na_2(CO_3)$$

Thus, nothing would happen.

There are other types of reactions that occur in chemistry.

Combination	$A + B \rightarrow AB$
Decomposition	$AB \rightarrow A + B$
Single Replacement	$A + BX \rightarrow AX + B$
Double Replacement	$AX + BY \rightarrow BX + AY$

Examples of these types of reactions can be found all around us.

Combination	$A + B \rightarrow AB$
Rusting of Iron	$4Fe + 3O_2 \rightarrow 2Fe_2O_3$
Decomposition	$AB \rightarrow A + B$
Production of Quick Lime	$Ca(OH)_2 \rightarrow CaO + H_2O$
Single Replacement	$A + BX \rightarrow AX + B$
Polishing of Silver by Soaking with Aluminum	$2Al + 3Ag_2S \rightarrow 6Ag + Al_2S_3$
Double Replacement	$AX + BY \rightarrow BX + AY$
Extraction of Magnesium from Seawater	$MgCl_2 + 2NaOH \rightarrow 2NaCl + Mg(OH)_2$

But, there are three additional types of common reactions – combustion, redox (oxidation-reduction) and acid/base reactions. A combustion reaction always involves oxygen. In these types of reactions, you have a fuel source, typically an organic molecule like methane or propane, which when combined with oxygen produces carbon dioxide, water and heat. Combustion reactions are used all the time as a means of releasing energy to allow for work to happen. In a redox (oxidation-reduction) reaction, electrons are exchanged between compounds. Redox reactions are involved in photosynthesis and batteries. Acid/base reactions or neutralization reactions involve the exchange of protons or hydrogen atoms and result in the production of water (H20). While these three common reactions have special names, they are essentially a special cases of the combination, decomposition, single replacement and double replacement reactions.

So in this laboratory, you are going to review, predict, and test, these different types of reactions.

General Safety Considerations for this Activity

During this laboratory, you will be using many of the same materials you have used previously – washing soda, calcium chloride, table salt, and Epsom salt. Thus, you already are familiar with the hazards associated with these materials. Two new chemicals will be introduced in this laboratory, steel wool and vinegar. Again as they are common materials, these do not necessarily present an unreasonable hazard.

During the laboratory, you will be combining different chemicals. And as previously discussed, it is very important to know what the possible implications are for mixing different chemicals, as their reaction products could produce harmful gases. However, the chemicals chosen for this laboratory are not anticipated to produce a hazardous reaction. It is important that you carefully review the procedures prior to conducting the laboratory and complete your hazard analysis. Gloves and goggles should be worn while conducting the laboratory. Care should also be taken not to inhale any of the products produced from the reactions.

Equipment required for the Activity:

- Vinegar
- Calcium chloride
- Washing Soda
- Epsom salt
- Table Salt
- Steel wool
- Tongs
- Lighter
- Test tubes
- Distilled water
- Test tube holder

Objectives:

- Writing chemical equations
- Predicting reactions
- Determining the reaction type

Procedure:

Obtain a SDS for steel wool and vinegar. Record information related to these compounds in your laboratory notebook. Recall, the information you have been collecting includes:

- The common name
- Chemical name
- Common use
- Any special handling instructions

- The trade name or product specific name
- Chemical formula
- Warnings/Cautions

Step by step procedures are to follow; however, an overview and some pre-laboratory work must be completed prior to beginning the experimental work. In this laboratory, you will perform the following experiments:

a) Adding Vinegar to a test tube prepared with a small amount of solid
material. For this experiment you will use table salt, Epsom Salt, calcium chloride, and washing soda.

b) You will make an ionic solutions of table salt, calcium chloride, Epsom salt and washing soda, and combined them in various combinations to determine if a reaction occurs.

c) You will ignite the steel wool using the lighter.

Pre-Laboratory Work

Part A – Adding vinegar to small amounts of solid material.

1) Determine the chemical formulas of the each of the reactants; vinegar, table salt, Epsom Salt, calcium chloride, and washing soda.

2) In this portion of the laboratory, you will have a small amount of solid in each test tube. You will be adding vinegar to the test tube and observe the reaction. As discussed in the safety section, you need to know the potential products of each reaction. Thus, you need to write down the potential chemical equation for each reaction. Write a balanced equation for the anticipated reactions. (Note: you should have four chemical equations; one for vinegar plus table salt, one for vinegar plus Epsom Salt, etc.)

3) Using the information provided in the introduction to the laboratory, predict the possible products of the reaction, and predict whether or not a reaction will occur.

Procedure (Continued):

Part B - Combining Ionic Solutions

1) In this portion of the laboratory, you will be mixing ionic solutions. You are going to use the following combinations:

> Combine the ionic solution of table salt with the ionic solution of Epsom Salt
>
> Combine the ionic solution of table salt with the ionic solution of calcium chloride
>
> Combine the ionic solution of table salt with the ionic solution of washing soda
>
> Combine the ionic solution of Epsom salt with the ionic solution of calcium chloride
>
> Combine the ionic solution of calcium chloride with the ionic solution of washing soda

Write a balanced equation for each of the anticipated reactions.

2) Using the information provided and your solubility tables, predict whether or not the reaction will occur.

Part C - Steel Wool

You are going to ignite the steel wool, hence you are going to introduce oxygen to the iron metal. Write the predicted reaction.

Laboratory Procedure

1. Having completed your pre-laboratory work, complete your hazard analysis. Document the hazard analysis in your laboratory notebook.

Part A – Adding vinegar to small amounts of solid material.

2. Obtain 4 clean dry test tubes.

3. Label the test tubes 1, 2, 3, and 4.

Laboratory #13
Gas Producing Reactions

Introduction

In your last laboratory, you had a chance to observe several types of chemical reactions. When you added the vinegar to the solids, you should have observed some bubbling when you reacted the vinegar with the washing soda. The bubbling occurred because one of the products of the reaction was a gas. Let's look at this reaction a bit closer

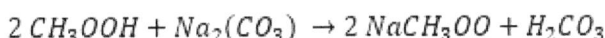

$$2\,CH_3OOH + Na_2(CO_3) \rightarrow 2\,NaCH_3OO + H_2CO_3$$

But, the products written, don't appear to be gases. So, why did you see a gas evolve? Because, one of the products is unstable – the H_2CO_3 decomposes spontaneously into water and carbon dioxide, or

$$H_2CO_3 \rightarrow H_2O + CO_2$$

The carbon dioxide is a gas and it is what caused the bubbles that you observed. There are lots of gas producing reactions. You even learned about one, when you were looking at the hazards associated with bleach, when mixed with the wrong materials chlorine gas can be evolved. One has to be very careful when working with gas producing reactions because the gas that may be produced can be harmful hazardous. This is why it is important to know about the materials you are working with and know what the potential reactions that may occur. (This is also why in the last laboratory, you needed to write out your chemical equations prior to performing the reactions.)

As you learned earlier, chlorine gas is harmful. But, oxygen and carbon dioxide may be harmful in a different way. If your reaction produces an oxygen gas, you need to be careful because the introduction of oxygen may result in a combustion reaction, fire. If your reaction produces a lot of carbon dioxide in a confined space, there may be a suffocation hazard. Thus, understanding gas producing reactions is important not only from a chemistry aspect but from a safety aspect as well. This is why most chemistry laboratories have fume hoods. A fume hood is a safety device designed to allow the scientist to perform reactions without being exposed to the harmful gas.

Common reactions that produce gases include:

Combustion reactions – products are carbon dioxide and water

A metal reacting with an acid – produces hydrogen gas

Acid/Base reactions – produce a salt, water, and carbon dioxide

Decomposition reactions – the produced gas is dependent on the material

In this laboratory, you are going to explore two different types of gas producing reactions – an acid/base reaction and a decomposition reaction. Due to the small quantities and types of gases produced, these are safe to conduct in your kitchen. You are even going to be able to verify the type of gas produced. You may already be familiar with both types of reactions. So, let's get started.

General Safety Considerations for this Activity

In this laboratory, you are going to produce a gas from the reaction. The gases to be produced are oxygen and carbon dioxide. In the introduction, hazards associated with oxygen and carbon dioxide were discussed. However, the amount of these materials that will be produced is very small, and you will be working in a large open area. Remember, it is a good safety practice not to smell the products of any reaction.

You will be using a chemical that you haven't used previously, hydrogen peroxide. Be sure to obtain the SDS and review the hazards associated with the material. The solution you will be using is a 3% solution of hydrogen peroxide and so should not be particularly hazardous, but it still needs to be used with caution. Additionally, this laboratory uses baking soda rather than washing soda. If you haven't obtained a SDS for baking soda you should do so and review the potential hazards associated with this material as well.

You will be using an open flame (from a wooden match), so care should be taken to ensure that there is no fire hazard.

As with all of your laboratory activities, gloves and goggles should be worn while conducting the experiment.

Equipment required for the Activity:

- 3% hydrogen peroxide solution
- Distilled vinegar
- Baking soda
- Yeast
- Two test tubes
- Test tube rack
- Wooden matches
- 10 mL graduated cylinder

Objectives:

- Observation of gaseous reactions

Procedure:

1. Obtain a copy of the safety data sheet for hydrogen peroxide and baking soda. Record information related to these materials in your laboratory notebook. Recall, the information you have been recording includes:

- The common name
- Chemical name
- Common use
- Any special handling instructions

- The trade name or product specific name
- Chemical formula
- Warnings/Cautions

2. Read through the entire laboratory before starting. Complete your hazard analysis, and document this in your laboratory notebook.

3. Gather the materials to conduct the experiment.

4. You are going to perform two reactions – mixing the baking soda and vinegar, and adding yeast to the hydrogen peroxide. Write the chemical equation for the baking soda and vinegar reaction, recall that the H_2CO_3 will decompose into water and carbon dioxide.

Procedure (Continued):

The reaction that occurs with the hydrogen peroxide is a decomposition reaction catalyzed by the yeast. This means that the yeast is not a reactant, but allows the reaction to proceed. See the catalyst box. Write out the decomposition reaction for hydrogen peroxide knowing that the products of the reaction are water and oxygen gas.

Catalyst

A catalyst is a substance that causes a reaction to happen or accelerates a chemical reaction without itself being affected.

Your gas detector is going to be a lit match. What is going to happen when the match is exposed to carbon dioxide? What is going to happen when the match is exposed to more oxygen? You will need to have a lit match ready when you perform the reactions.

5. Place approximately 1 gram of baking soda into one of the test tubes. Measure out 1 mL of distilled vinegar.

6. Prepare your gas detector (light the match). Pour the vinegar into the test tube with the baking soda. Move the lit match over the test tube while the reaction is occurring. Observe. Dispose/extinguish the match properly. Touch the test tube near where the reaction occurred.

7. Document your observations.

8. Place approximately 1 gram of yeast into a test tube. Measure out 1 mL of the hydrogen peroxide.

9. Prepare your gas detector (light the match). Pour the hydrogen peroxide into the test tube with the yeast. Move the lit match over the test tube while the reaction is occurring. Observe. Dispose/extinguish the match properly. Touch the test tube near where the reaction occurred.

10. Document your observations.

11. Dispose of the reaction contents down the sink and clean-up work area.

Summing It Up:

In this laboratory, you were able to observe two different types of gas producing reactions. You were also introduced to other types of gas producing reactions and introduced to a reaction aid called a catalyst. You should have also observed another product of a reaction, heat.

Laboratory #14
Putting What You Have
Learned Together

Introduction

Throughout this course, you have been building your laboratory skills. You have learned about physical properties, how to obtain chemical information, and have started exploring chemical reactions. You have also learned about the scientific method, how to document your results, and how this might help you develop your own experiments. Now, it is time to pull all of this together, so that you can see how these skills might have a practical application.

For this laboratory, you are going to develop your own procedure to help you to determine which white powder is which. There is a bit of cheating involved because you know what powders you are starting with and you know what reagents you have at your disposal. But, the objective of this laboratory is for you to develop your procedure, so that if someone else where to have to determine which powder was which they could follow your method and get the correct answer.

So, let's set up a scenario:

> *A person is putting together a kit to take to a school for a demonstration. Each kit contains a jar of washing soda, crushed chalk and table salt. All three are white powders. Before labeling the jars, the person has put everything away and gets called away for a few hours. The person returns only to discover, that someone has rearranged the jars. Now, the person doesn't know which jar is which.*

How is the person going to determine which jar contains which white powder? The only other materials available to help with the problem is distilled water and vinegar. Your task for this laboratory is to develop a procedure using the distilled water and vinegar to determine which white powder is the washing soda, crushed chalk and table salt.

Your task for this laboratory is to develop a procedure using the distilled water and vinegar to determine which white powder is the washing soda, crushed chalk and table salt.

Here are some things to think about when developing your procedure:

1) You have done a laboratory investigating solubility in water.

2) You have done a laboratory looking at reactions involving water and vinegar.

3) You have done a laboratory observing gas production.

General Safety Considerations for this Activity

During this laboratory, you will be working with materials that you have used before. Be sure to develop your hazard analysis and wear your gloves and goggles when performing the laboratory.

Equipment required for the Activity:

- Washing soda
- Distilled water
- Vinegar
- Table Salt
- Crushed chalk (Calcium carbonate)
- Any other equipment that you determine you need.

Objectives:

- Using previous work to develop a new procedure.

Procedure:

1. Do your hazard analysis.

2. Develop your procedure based on the following scenario:

> *A person is putting together a kit to take to a school for a demonstration. Each kit contains a jar of washing soda, crushed chalk and table salt. All three are white powders. Before labeling the jars, the person has put everything away and gets called away for a few hours. The person returns only to discover, that someone has rearranged the jars. Now, the person doesn't know which jar is which.*
>
> *How is the person going to determine which jar contains which white powder? The only other materials available to help with the problem is distilled water and vinegar.*

3. Document the procedure in your laboratory notebook.

4. Conduct your experiment to determine if your procedure worked. Could you verify which material was which based no your procedure? If not, what would you change and try again.

5. Write a laboratory report outlining your procedure and findings.

Equipment List

	Required in Laboratory
Safety Items	
Goggles/Glasses	4, 5, 6, 8, 10, 11, 12, 13, 14
Gloves	4, 5, 6, 8, 10, 11, 12, 13, 14
Fire Extinguisher	Not specifically required but recommended to have available.
Laboratory Equipment	
Graduated Cylinder - 10 ml	3, 5, 10, 11, 13
Graduated Cylinder - 100 ml	3, 5
Heat Source	6, 10
Measuring Cups	3, 8
Sauce Pan or Beakers for water bath	6, 10
Scale	3, 4, 5, 7, 8, 10, 11
Test Tubes	10, 11, 12, 13
Test Tube Rack	10, 11, 12, 13
Test Tube Tongs	10
Thermometer	6, 8, 10
Chemicals	
3% Hydrogen Peroxide Solution	13
Baking Soda	5, 13
Calcium Chloride	10, 12
Crushed Chalk	14
Epsom Salts	11, 12
Ice	6, 8
Paraffin Wax	6
Steel Wool	12
Table Salt (non-iodized)	5, 6, 8, 12, 14
Vinegar	12, 13, 14
Washing Soda	8, 11, 14
Water (tap and distilled)	3, 5, 6, 8, 10, 11, 12, 14
Yeast	13

Other Supplies

Supply	Required in Laboratory
Calculator	All
Coffee Filter	11
Composition Book	All except Laboratory 1
Disposable tart tins	6
Hot Pads	6
Internet Access	All
Kitchen Tongs	6, 12
Lighter	12
Marker	10
Nail or screw	3
Plain copy paper	3
Plastic Spoons	10
Ruler (in centimeters and inches)	3
Scissors	11
Set of bolts, washers, and nuts (Experiment requires 8 bolts, 8 nuts and 16 washers)	7
Small block of wood	3
Stirring Rod (glass rod if available or chop stick)	6, 10, 11
Styrofoam cups (2)	8
Teaspoon	5
Ten post-1983 pennies	3, 4, 5
Ten pre-1983 pennies	5
Wax paper or foil for measuring out chemicals	10, 11
Wooden Matches	13

www.ingramcontent.com/pod-product-compliance
Lightning Source LLC
Chambersburg PA
CBHW051118200326
41518CB00016B/2545